THE QUALITY PROMISE

QUALITY AND RELIABILITY

A Series Edited by

Edward G. Schilling

Center for Quality and Applied Statistics
Rochester Institute of Technology
Rochester, New York

1. Designing for Minimal Maintenance Expense: The Practical Application of Reliability and Maintainability, *Marvin A. Moss*
2. Quality Control for Profit, Second Edition, Revised and Expanded, *Ronald H. Lester, Norbert L. Enrick, and Harry E. Mottley, Jr.*
3. QCPAC: Statistical Quality Control on the IBM PC, *Steven M. Zimmerman and Leo M. Conrad*
4. Quality by Experimental Design, *Thomas B. Barker*
5. Applications of Quality Control in the Service Industry, *A. C. Rosander*
6. Integrated Product Testing and Evaluating: A Systems Approach to Improve Reliability and Quality, Revised Edition, *Harold L. Gilmore and Herbert C. Schwartz*
7. Quality Management Handbook, *edited by Loren Walsh, Ralph Wurster, and Raymond J. Kimber*
8. Statistical Process Control: A Guide for Implementation, *Roger W. Berger and Thomas Hart*
9. Quality Circles: Selected Readings, *edited by Roger W. Berger and David L. Shores*
10. Quality and Productivity for Bankers and Financial Managers, *William J. Latzko*
11. Poor-Quality Cost, *H. James Harrington*
12. Human Resources Management, *edited by Jill P. Kern, John J. Riley, and Louis N. Jones*
13. The Good and the Bad News About Quality, *Edward M. Schrock and Henry L. Lefevre*
14. Engineering Design for Producibility and Reliability, *John W. Priest*
15. Statistical Process Control in Automated Manufacturing, *J. Bert Keats and Norma Faris Hubele*

16. Automated Inspection and Quality Assurance, *Stanley L. Robinson and Richard K. Miller*
17. Defect Prevention: Use of Simple Statistical Tools, *Victor E. Kane*
18. Defect Prevention: Use of Simple Statistical Tools, Solutions Manual, *Victor E. Kane*
19. Purchasing and Quality, *Max McRobb*
20. Specification Writing and Management, *Max McRobb*
21. Quality Function Deployment: A Practitioner's Approach, *James L. Bossert*
22. The Quality Promise, *Lester Jay Wollschlaeger*

THE QUALITY PROMISE

Lester Jay Wollschlaeger
Quality Innovators, Inc.
Corvallis, Oregon

Marcel Dekker, Inc. New York • Basel • Hong Kong
ASQC Quality Press Milwaukee

Library of Congress Cataloging-in-Publication Data

Wollschlaeger, Lester Jay
The quality promise / Lester Jay Wollschlaeger
p. cm. - - (Quality and reliability; 21)
Includes bibliographical references and index.
ISBN 0-8247-8389-1
1. Quality control. 2. Reliability (Engineering) I. Title
II. Series
TS156.W65 1990
658.5'62- -dc20 90-49146
CIP

This book is printed on acid-free paper.

Marcel Dekker, Inc.
270 Madison Avenue, New York, New York 10016

American Society for Quality Control
310 West Wisconsin Avenue, Milwaukee, Wisconsin 53203

Current printing (last digit):
10 9 8 7 6 5 4 3 2 1

Printed in the United States of America

Dedicated with love to Diane
who believes before I do,
who encourages me when I need it most,
and who understands me and loves me anyway.

About the Series

The genesis of modern methods of quality and reliability will be found in a simple memo dated May 16, 1924, in which Walter A. Shewhart proposed the control chart for the analysis of inspection data. This led to a broadening of the concept of inspection from emphasis on detection and correction of defective material to control of quality through analysis and prevention of quality problems. Subsequent concern for product performance in the hands of the user stimulated development of the systems and techniques of reliability. Emphasis on the consumer as the ultimate judge of quality serves as the catalyst to bring about the integration of the methodology of quality with that of reliability. Thus, the innovations that came out of the control chart spawned a philosophy of control of quality and reliability that has come to include not only the methodology of the statistical sciences and engineering, but also the use of appropriate management methods together with various motivational procedures in a concerted effort dedicated to quality improvement.

This series is intended to provide a vehicle to foster interaction of the elements of the modern approach to quality, including statistical applications, quality and reliability engineering, management, and motivational aspects. It is a forum in which the subject matter of these various areas can be brought together to allow for effective integration of appropriate techniques. This will promote the true benefit of each, which can be achieved only through their interaction. In this sense, the whole of quality and reliability is greater than the sum of its parts, as each element augments the others.

The contributors to this series have been encouraged to discuss fundamental concepts as well as methodology, technology, and procedures at the leading edge of the discipline. Thus, new concepts are placed in proper perspective in these evolving disciplines. The series is intended for those in manufacturing, engineering, and marketing and management, as well as the consuming public, all of whom have an interest and stake in the improvement and maintenance of quality and reliability in the products and services that are the lifeblood of the economic system.

The modern approach to quality and reliability concerns excellence: excellence when the product is designed, excellence when the product is made, excellence as the product is used, and excellence throughout its lifetime. But excellence does not result without effort, and products and services of superior quality and reliability require an appropriate combination of statistical, engineering, management, and motivational effort. This effort can be directed for maximum benefit only in light of timely knowledge of approaches and methods that have been developed and are available in these areas of expertise. Within the volumes of this series, the reader will find the means to create, control, correct, and improve quality and reliability in ways that are cost effective, that enhance productivity, and that create a motivational atmosphere that is harmonious and constructive. It is dedicated to that end and to the readers whose study of quality and reliability will lead to greater understanding of their products, their processes, their workplaces, and themselves.

Edward G. Schilling

Preface

The Quality Promise: "A real understanding of quality captures the system, tames it and puts it to work for one's personal use."*

There is a tendency in business to focus on the big picture. When I worked in Quality Assurance Programs the big picture emphasized systems for assuring product quality. It focused on establishing centralized systems to reduce scrap and rework, and to decrease the chance of a customer receiving a defective product. Gradually I came to realize that this big picture is actually made up of smaller pictures.

*The full quote reads, "A real understanding of quality doesn't just serve the system, or even beat it, or even escape it. A real understanding of quality captures the system, tames it, and puts it to work for one's personal use, while leaving one completely free to fulfill his inner destiny." [From *Zen and the Art of Motorcycle Maintenance* by Robert Pirsig, New York: Morrow (1979).]

The focus of this book is the smallest, but perhaps the most important one within that system: you, the individual worker. The goal is not to develop a centralized quality system for managing quality on a grand scale. Rather it is the more modest one of focusing on just one individual within that system, and having that one individual operate with excellence. America has relied on individual integrity and self-reliance in the past and needs to emphasize it again. We do need quality systems, but those systems must be rebuilt on a foundation that recognizes the individual.

This book is about the individual at the foundation of the quality system, the quality master. It explores the beliefs and expectations of this role within the quality system. It also creates the linkages to explain the quality promise: that a real understanding of quality doesn't just serve the system, but captures it and puts it to your own personal use.

If this book creates just one quality master, its objective has been met.

Lester Jay Wollschlaeger

Contents

THE QUALITY PROMISE

Part I

BELIEFS

1

Examining Beliefs

RECOGNIZING BELIEFS

Would you believe me if I were to tell you that Ed McMahon is going to call you about the American Publishers' Sweepstakes next week? Probably not. But if you were to win, would you believe that the Internal Revenue Service would contact you before you could write the deposit in your checkbook? Sure. These are examples of beliefs we may have. Beliefs are the mental acceptance of the truth as we know it.

Most of the time I recognize my beliefs. For example, there are things that I believe I do well. There are other things that I believe I do not do well. I have always enjoyed mathematics and have developed a belief that I am good at this. Drawing, on the other hand, has always been difficult for me, and I do not feel that I am very good at drawing—at least at this time.

I also have beliefs that I do not recognize. For example, five years ago I was asked to be a member of a panel to discuss the emerging importance of quality. This was a topic that had always interested me, and it was my first chance to be on television.

I prepared for the panel discussion and anxiously awaited my first television appearance. I was told to have any graphics that I would need prepared and submitted a week in advance. I had them ready two weeks in advance. I was told to wear a dark suit. I bought a dark gray pinstripe suit right out of *Dress for Success.* I was told to arrive at 5:00 P.M. I arrived at 4:30 P.M. I was ready for my grand debut.

About 5:00 P.M. a woman came over and asked when I was to be on. "Six o'clock," I said. "Then follow me," she said. I followed her to a small room where she told me to sit down. Then came something I did not expect. She put what I thought was a barber's apron on me. Then she brought out what appeared to be women's makeup.

It was then that I discovered a (hidden) belief that I had—that real men do not wear makeup. Never in my life had I ever wanted to wear (women's) makeup. It seemed totally inappropriate now that I was to appear before thousands of people that I would put on (women's) makeup. It didn't matter to me whether real men did wear makeup when they went on television. What was going through my mind was how to get out of the chair with my perceived masculinity still intact. My beliefs, not the truth, caused me to act this way. I acted in accordance with the truth *as I believed it to be.*

SOME FALSE BELIEFS CAN LIMIT OUR ACCOMPLISHMENTS

I am not the only one to act in accordance with beliefs that are false. Entire nations have done this. One of the most famous examples of this occurred in the days of Columbus. One belief at that time was that the world was flat. Mapmakers and sailors alike drew the world on paper and in their minds as being flat. They believed that if you sailed out too far, you would fall off the edge. Columbus had a different belief. He believed that the world was round and that you could sail around it.

The effect of believing that the world was flat limited the possibilities of what could be achieved. By having a different belief about the world, Columbus was able to open up new navigational possibilities. He enabled mankind to use the existing skills of seacraft for vastly greater undertakings.

We, too, may have beliefs that limit us—we may have flat worlds. It could be a belief that real men do not wear makeup, which can limit how well we appear on television. It could be a belief that we cannot make a difference in our company. It could be a belief that incomplete instruction on how to perform a job correctly is acceptable and that if we ask a question we will be judged incompetent. It could be a fear that if we point out an error to a coworker, we will be considered too picky and not a team player.

EVALUATING BELIEFS

Sensitive or soft-hearted; persistent or stubborn; curious or nosy; exacting or critical; determined or pushy; persuasive or manipulative. Words expressing different beliefs about people can describe similar traits. The difference is sometimes a matter of degree. Other times the appropriateness of the beliefs could be determined by (1) the goals to be achieved; (2) the environment, the culture, or the point of view; or (3) the usefulness of the belief to explain a phenomenon.

APPROPRIATENESS OF THE BELIEF MAY DEPEND ON THE GOAL

A Chinese farmer once inherited a horse. A neighbor came by and told the farmer, "You are very fortunate to have received this horse." "Maybe," said the farmer. The son of the farmer was riding the horse, fell off, and broke his leg. The neighbor once again came by and now said, "How unfortunate it was that your son broke his leg." "Maybe," said the farmer. The next day, the army came to recruit all the young men that could fight. Once again the neighbor came by. "You are very fortunate that your son was not recruited," said the neighbor. "Maybe," the farmer replied.

Shakespeare said it a little differently: "There is nothing either good or bad, but thinking makes it so" (Hamlet II, ii). Similarly, beliefs are not necessarily good or bad until we set a goal. Then we must ask, "Does this belief lead me toward achieving the goal, or does it lead me away?" If it leads me toward my goal then it is an appropriate belief. If it leads me away, then the belief is inappropriate.

For example, is being independent good or bad? Some people might say it is good. Standing on one's own feet, or independently making conclusive decisions in the face of adversity, are positive aspects of independence. Yet there are times when independence can be viewed as counterproductive. Consider a basketball team, for example. In a team setting, being independent, doing one's own thing, prevents the teamwork necessary to win. If the goal is for the team to win, being independent will lead away from the goal and will be considered inappropriate. So the appropriateness of some beliefs cannot be determined until you set a goal.

APPROPRIATENESS MAY DEPEND ON THE CULTURE

Practices that may be considered bad in one culture may be perceived as perfectly acceptable in another. Nepotism, the practice of giving a job to a relative, may be considered of dubious morality in one culture, in today's United States, for instance. In other cultures, a traditional Chinese one, for example, it may be the very essence of ethical behavior, in that it satisfies one's moral obligations to one's family.

OPPOSING BELIEFS MAY BOTH BE USEFUL

There are also times where two totally opposing beliefs are both appropriate depending on the phenomena to be explained. For instance, in physics there have been two opposing beliefs about the nature of subatomic units. It is called the wave-particle duality. One belief is that a subatomic unit like an electron is by nature a particle. The opposing belief is that the subatomic unit is by nature a wave.

Both beliefs have been supported by experiments. Light, for example, travels through space showing all the characteristic behavior of a wave. Light shows wave properties when diffracted through a prism. A rainbow is formed when light is broken up into different frequencies, which appear as different colors.

On the other hand, interpretation of light as particles is substantiated by the so-called photoelectric effect. Light is emitted and absorbed in the form of quanta or photons (i.e., as particles). This is demonstrated when ultraviolet light is shone on the surface of certain metals. The light can kick out electrons from the surface of the metal. The interpretation here is that light must consist of moving particles. One set of experiments can be explained only through the wave property of light, but the other set of experiments can be explained only by collisions of particles with electrons.

Physicists used these differences to formulate quantum theory, which is based on the idea that the particle picture and the wave picture are two complementary descriptions of the same reality. Each of them is needed to give a full description of the complete reality, and both are to be applied within the limitations of the model.

This notion of complementarity has become an essential part of the way physicists think about nature. Niels Bohr, a Nobel prize winner for his description of the atomic structure, has often suggested that it might be a useful concept outside the field of physics.

SUMMARY

A belief is the mental acceptance of a truth as we believe it to be. A belief has a powerful effect on our lives, since we do not act in accordance with the truth, but in accordance with our beliefs of the truth.

The goodness of a belief is the effect it has on us. Beliefs are not right or wrong but rather appropriate or inappropriate. The appropriateness may depend on goals, environment, culture, or viewpoint. In fact there are times when opposing beliefs may both be useful.

2

Beliefs About Quality

We have beliefs about quality as about other things. Some of these beliefs are hidden. Some are complementary. Some are true and we act in accordance with them. Some are not true and we act in accordance with them as well.

Some of the differences in beliefs can be explained by different aspects of quality. The American Society for Quality Control poster shown in Figure 2.1 shows some of the many synonyms of quality.

ONE WORD, TWO (INDEPENDENT) DIMENSIONS

The fact that quality has at least two independent dimensions was made very clear to me about eight years ago. I was working for a boss called

A systematic approach to the search for excellence. (Synonyms: productivity, cost reduction, schedule performance, sales, customer satisfaction, teamwork, the bottom line.)

Figure 2.1 The American Society for Quality Control poster.

Big Al. Big Al got his name from his intimidating size. He was a huge former tight end for the Houston Oilers. He was equally intimidating mentally with two degrees from Stanford, one in physics, the other a Master's in Business Administration.

Big Al had worked in Germany and recognized that my last name was German. One day he called me into his office. "Wie geht's?" (How are you doing?) he said, knowing that about the only German word I know is Wollschlaeger. He then proceeded, "Les, you've been with us about a month, I think it is about time that you become bilingual."

Thoughts raced through my mind of trips to Europe. What a nice perk! Big Al then said, "The first word I would like you to learn is 'quality'. It is the uniformity, the consistency, of the products, and the product's conformance to specifications. Okay?"

"Sure," I responded, although I wasn't sure where he was headed.

"The second word I would like you to learn is 'quality'." It sounded as if he were about the repeat himself. "Quality is the perceived grade of the product in the marketplace." At first, I thought he was proving that he could speak out of both sides of his mouth at one time, but he was emphasizing that one word, *quality,* has two distinct, independent meanings.

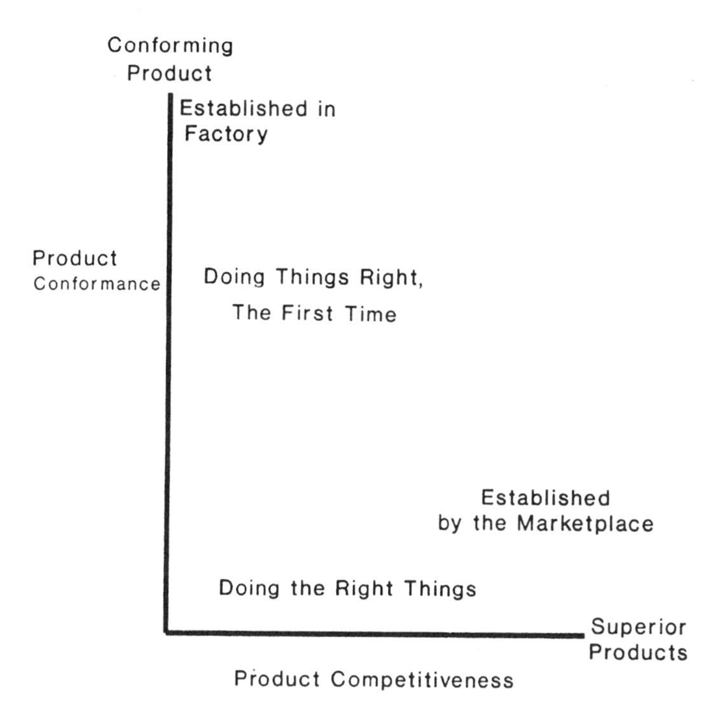

Figure 2.2 Two dimensions of quality.

Big Al then continued: "We need to be bilingual. Do you think everyone can learn to speak bilingually?" Then in my best broken German I answered, "Oh, yaaaah."

Figure 2.2 illustrates the two dimensions of quality. In one direction is the conformance of the product to the requirements. It focuses on a uniform, consistent product meeting specifications. It is concerned about doing things right the first time. The place where the quality of conformance becomes an issue is typically in the factory.

The second dimension of quality occurs in the marketplace. In this dimension the issue is the perceived grade of the product. It is concerned with the competitive position of the product relative to other products that are being sold. It is concerned with doing the right things.

THREE BELIEFS ABOUT QUALITY

Some beliefs about quality have been identified, studied, and in some cases quantified. Three beliefs about quality are the following:

1. The quality of conformance decreases the cost.
2. Perceived quality increases the selling price.
3. Most errors occur because the criteria for self control are not met.

Belief 1: The Quality of Conformance Decreases the Cost

The first dimension of quality is the quality of conformance. Improving the quality (i.e., uniformity, consistency, or predictability) generally reduces the cost. Products that do not have to be reworked or scrapped can be made more efficiently.

The American Society for Quality Control has standardized the quality cost structure. The categories are listed here.*

1. *Prevention:* The cost associated with personnel engaged in designing, implementing, and maintaining the quality system. Maintaining the quality system includes auditing the system.
2. *Appraisal:* The costs associated with the measuring, evaluating, or auditing of products, components, and purchased materials to assure conformance with quality standards and performance requirements.
3. *Internal failures:* The costs associated with defective products, components, and materials that fail to meet quality requirements and result in manufacturing losses.
4. *External failures:* The costs which are generated because of defective products being shipped to customers.

* ASQC Quality Cost-Cost Effectiveness Technical Committee, *Quality Costs–What and How, 2nd Ed.*, ASQC, Madison, Wisconsin, 1971.

The committee recommends that systems be established to record costs according to the following detailed categories:

Prevention

1. Quality planning and process control planning
   a. Quality planning—quality control engineering type work
   b. Process quality control—that portion of compensation and costs associated with implementing the quality plans and procedures
2. Design and development of quality measurement and control equipment
3. Quality planning by functions other than quality control
4. Quality training
5. Other prevention expense

Appraisal

1. Receiving or incoming test and inspection
2. Laboratory acceptance testing
3. Inspection and test
4. Checking labor
5. Setup for inspection and test
6. Inspection and test material
7. Quality audits
8. Outside endorsements or approvals
9. Maintenance and calibration of test and inspection equipment
10. Review of test and inspection data
11. Field testing
12. Internal testing and release
13. Evaluation of field stock and spare parts

Internal Failures

1. Scrap
2. Rework and repair
3. Troubleshooting

4. Reinspect, retest
5. Scrap and rework—fault of vendor
6. Material review activity
7. Downgrading

External Failures

1. Complaints
2. Product or customer service
3. Returned material processing
4. Returned material repair
5. Warranty replacement
6. Engineering error
7. Factory or installation error

These costs are shown in Figure 2.3.

Quality Cost and Profit

A useful aid for explaining the interrelation among the quality costs is the Veen Model shown in Table 2.1.

First Stage—No Special Action

In this model, the quality-related costs at the outset total 24 (arbitrary)

Table 2.1 The Veen Model for Quality-Related Costs

Cost Category	Stage 1	Stage 2	Stage 3	Stage 4
External Failures	20	3	2	1
Internal Failures	1	12	8	4
Sorting	1	3	2	1
Control	1	1	4	2
Prevention	1	1	1	2
Total	24	20	17	10

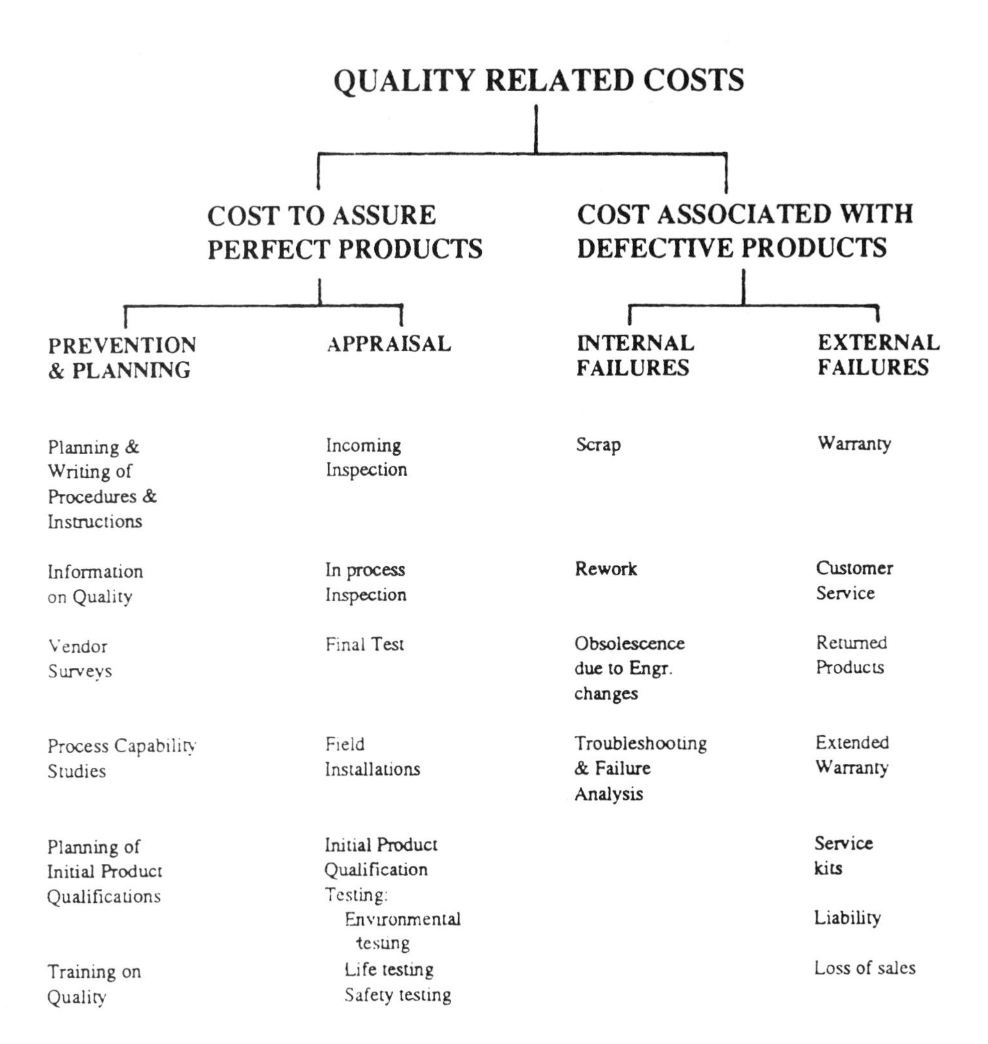

Figure 2.3 Breakdown of quality related costs.

units due primarily to defects reaching the users and causing high external failure costs. The high external costs are those costs associated with fixing products that have been shipped or the cost of having those products returned.

Second Stage—Sorting

The second stage occurs when the high external failures attract the attention of management. Inspection stations are established to prevent nonconforming products from being shipped to the customer. Very often this sorting operation is not 100% effective. If the sorting operation is 85% effective in eliminating the nonconforming, the cost of shipping defectives to the field decreases from 20 units to 3 units.

There are two costs in doing this. Inspection costs increase 3 units. Scrap costs, rework costs, and the cost associated with nonconforming material also increase. This results in 12 units of cost. The net gain in making this change is a decrease in cost from 24 units to 20 units.

Stage Three—Control

The major cost in this stage is the cost associated with scrap and rework. The focus here shifts from eliminating nonconforming products to reducing the internal failures. Management realizes that sorting is not a long term solution and that the process producing the nonconforming has to be controlled. Control charts and other techniques for listening to the process are introduced. The investment in control techniques to reduce scrap and rework is an additional 3 units. The benefit is a decrease in the internal failures from 12 to 8, a decreased need for sorting, reducing sorting costs from 3 to 2, and a further decrease in the proportion of defective units escaping to the field from 3 to 2.

Fourth Stage—Prevention and Total Quality Control

The next step is a more detailed and systematic investigation into the process for the nonconforming products; it includes improved process studies designed to remove the causes of the defects and increase the process capability. Greater attention is paid to improving the quality control systems. This investment into doing it right the first time is rewarded by further decreasing all other cost categories. Control costs are decreased because the source causing the problem is eliminated. Sorting costs, internal failure costs, and external failure costs are all further reduced because the nonconformity has less of a chance of occurring.

Studies Relating Failure Costs and Profits

Although quality-related cost is a good tool for thinking about the distribution of quality-related cost, there may be better indicators that correlate better profit. A study, by Francis X. Brown and Roger W. Kane at Westinghouse, indicates that failure costs (rather than total quality costs) may be better correlated with profit than other measures. This study, which was based on 64 business units, showed that every dollar decrease in failure costs, profits increased an average of $5.45. (See Figure 2.4.)

One possible explanation for this multiplier effect is that many of the costs of quality failures are not easily identified as failure costs, but that they affect profit anyway. Failures that cause shortages at subsequent operations increase production downtime. Some rework is absorbed in productivity and reported as labor variance. Chronic rework or excessive shrinkage may necessitate scheduling of overtime or purchase of additional production facilities. A completed product that fails in final test may result in increased work-in-process inventory and reduced billings for the month. Product failures in the field can, at least in the long run, contribute to reduced market share or poor price realization. They may also contribute to past-due or uncollectible receivables.

Belief 2: Perceived Quality Increases the Selling Price

The second dimension of quality is the grade of quality. Products may conform to the specifications, but there are differences in specifications. For example, there is a perceived difference in quality between a Chevrolet and a Cadillac although each may conform to its respective specifications.

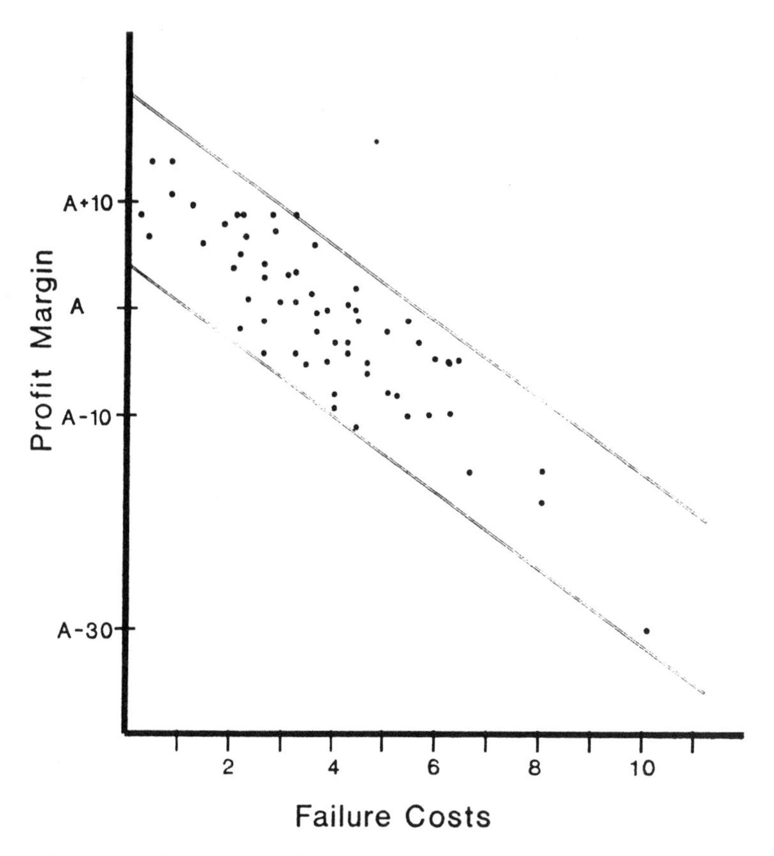

Figure 2.4 Profit vs. failure costs for 64 divisions of the Westinghouse corporation.

The higher selling price for a perceived higher quality can be summarized in a statement by Stanley Marcus, past chairman of Nieman- Marcus: “As long as oranges are two different sizes, people will pay more for the largest and juiciest.” This can be verified by taking a five-minute survey of food prices in your local grocery store. Compare the prices of the generic foods with those of the brand name products. The result of such a survey that I did is shown in Table 2.2. From this it appears that brand name products command about an 80% price premium.

This has also been researched and documented on a scientific basis. One of the first studies to try to identify and analyze the factors that impact profit was the Profit Impact of Marketing Strategies (PIMS) research. The PIMS team studied the effect of hundreds of variables with long-term financial performance. The analysis gave strong support to the proposition that market share is a major influence on profitability. It also identified other factors that had significant impact. One was perceived product quality. The score for the perceived product quality was appraised in the following terms: What was the percentage of sales of products or services from each business in each year that were superior to those of the competitor? What was the percentage of equivalent products? Inferior products? These factors included various quality and service delivery traits (other than price).

Table 2.2 Comparison Between the Selling Price of Generic and Brand Items

Item	Cost of generic brand	Cost of name brand
Tea bags	1.58	3.79
Nondairy coffee creamer	1.79	3.39
Laundry soap	1.89	3.89
Salad dressing	1.55	2.09
Oatmeal	1.69	1.99
Sweet peas	.39	.69
Sandwich bags	.59	1.25
Total	9.48	17.09

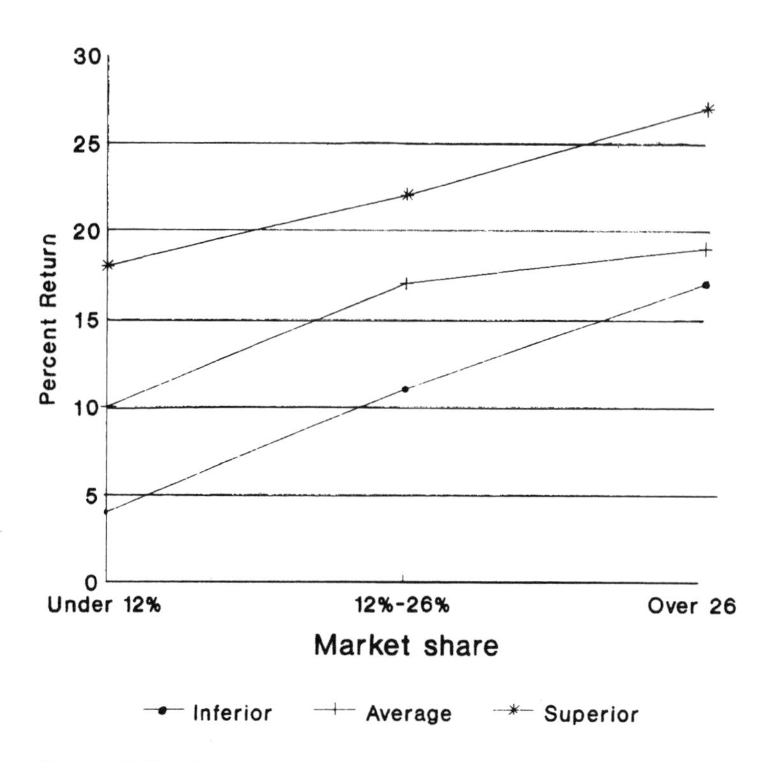

Figure 2.5 Effect of market share on return on investment.

The combined impact of market share and perceived product quality is shown in Figure 2.5. The best of all possible worlds is to have both high market share and superior quality. Businesses in this category averaged 28.3% return on investment. Even with a weak market position, however, superior quality producers earned 17.4%.

There have been other studies conducted since the initial PIMS research that are wholly consistent with the analyses of the PIMS team. *In Search of Excellence,* for example, researched 75 of the most financially successful companies. Of these 75, 5 were in the resource-extraction business. For those companies, efficiency may be an appropriate first principle. Of the remaining 70, 65 were successful by focusing on quality and service.

A similar study conducted by McKinsey and Co. for the American Business Conference focused on the management strategic practices among the 45 top performers (on a financial basis). The study showed that management practice in most of these companies was to provide higher value added products (often costing more to produce). In 43 of the 45 cases, the following was observed: "Winners compete by delivering a product that supplies superior value to customers."

A summary of the impact of quality is shown in Figure 2.6.

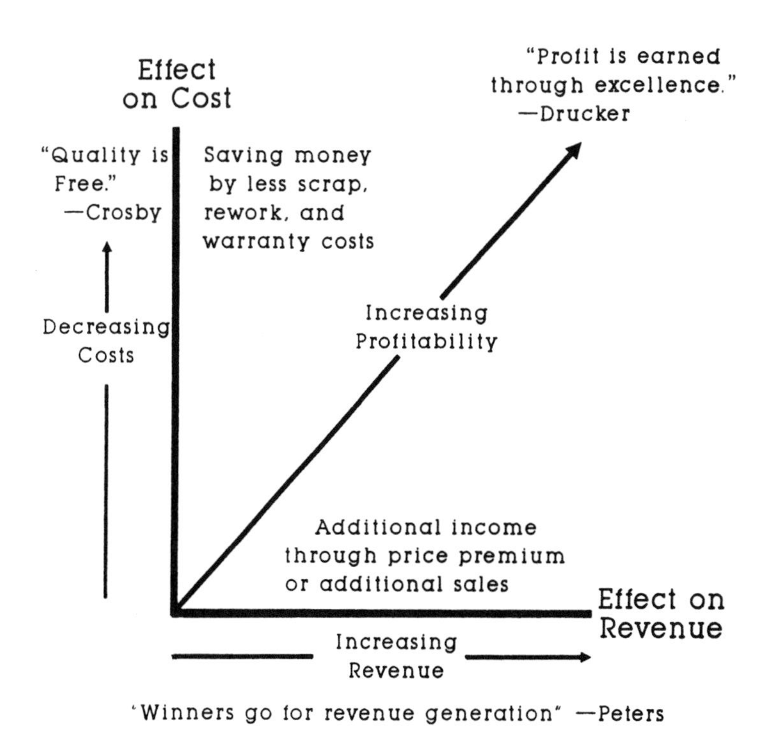

Figure 2.6 Financial impact of quality.

Belief 3: Most Errors Occur Because the Criteria for Self-Control Are Not Met

One belief has been that lack of motivation on the part of the worker is the cause of most errors. There have been, however, significant data published to refute this belief. Some of the earliest data were published by Dr. Joseph Juran. Dr. Juran uses a nonindustrial example, the golfer, to emphasize that motivation is not the issue.

> The golfer is obviously in a state of self control. He knows very well what he is supposed to do. He can observe with his senses what his actual performance is. He has tools identical to those used by experts. Why then are there so many golf balls in the lakes? Under conventional logic, the reason is that the golfer is not well motivated. Such a conclusion would be laughed at by anyone who plays golf or has observed golfers at play, since few people are as intensely motivated as golfers.*

In Dr. Juran's analysis of errors, the cause of 80% of the errors was not a lack of motivation, but a result of the worker not being in a state of self-control. The three criteria for self-control are

1. Knowledge of what is supposed to be done, i.e., the budgeted profit, the schedule, the product specification;
2. Knowledge of what is actually happening, i.e., the actual profit, the delivery rate, the extent of the conformance of the product to specifications; and
3. Means of regulating what is happening in the event that the goals are not being met. These means must always include the authority to regulate and the ability to regulate either by (1) varying the process under the person's authority or (2) varying the person's own conduct.

Dr. Juran's complete hierarchy of causes of errors is shown in Figure 2.7. In this analysis, Dr. Juran further subdivides the errors that occur when in a state of self-control into three categories. These are

*Juran, J.M., Frank M. Gryna, Jr., and R.S. Bingham, Jr., eds., *Quality Control Handbook,* Mc Graw-Hill, New York, 1951, p. 18-3.

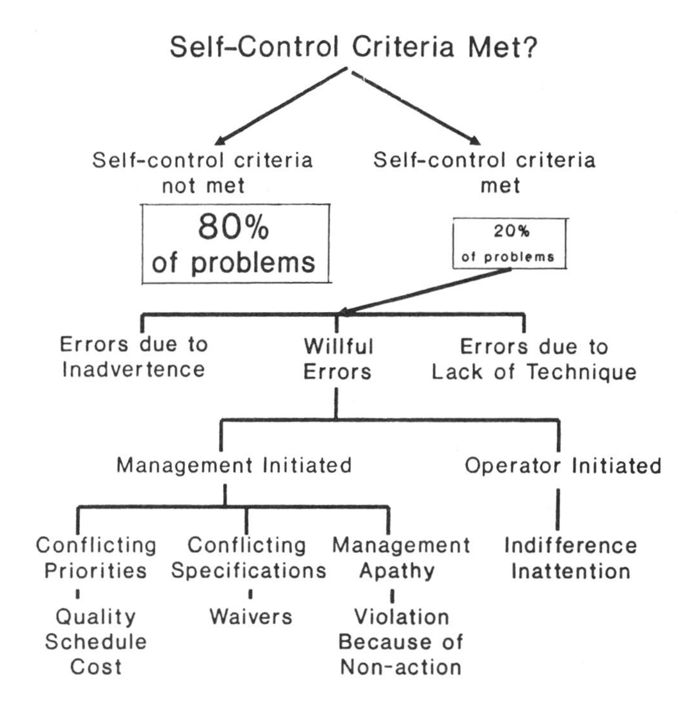

Figure 2.7 Juran's analysis of errors. Meeting the self-control criteria requires (1) knowledge of what to do, (2) feedback on how you are doing, and (3) the ability to correct differences.

1. Errors due to inadvertence. At the time of making the error, the person is unaware that he has made an error. He proofreads a memo and misses a spelling error or adds up a column of numbers and gets a wrong answer.
2. Errors due to lack of technique. These are errors that result from a lack of special knowledge or skill. Many times the special knowledge or knack may not be known even by the people who do not make the errors. The golfer is indeed an example.

3. Willful errors. Here the operator knows how to avoid making an error and despite this makes the error; he fails to meet the quality requirement.

At face value, willful errors seem unexcusable. Closer examination, however, makes clear that these errors arise from many reasons, good as well as bad.

One major category of willful errors could be considered management initiated. Three forms of the management initiated errors are caused by

1. Conflicting priorities. Management is faced with meeting multiple standards of performance: quality, schedule, cost, safety. Obviously life is easier if all standards are met, but if unforeseen events occur (deliveries to customers change), one of the standards may need to be compromised. One standard may have priority over the others.
2. Conflicting quality specifications. Management action in failing to enforce specifications also creates conflict, since repetitive actions become unofficial specifications. With no explanation, operators are led to believe that the unofficial specifications are acceptable.
3. Management apathy. In some cases managers fail to face up to quality problems encountered by operators; this forces the operators to find their own solutions.

For example, in one silicon wafer manufacturing plant, wafers were consistently too thick. Several studies were done to evaluate the variation in wafer thickness. Investigation showed that saw operators were purposely cutting thicker wafers because they were reprimanded if saw marks appeared after wafers were lapped and etched. They found that if they cut thicker wafers, the saw marks would be removed by subsequent processing and no reprimand would occur.

When the lapping and etching steps were adjusted to ensure adequate elimination of saw marks, the wafers started being cut to the right thickness.

SUMMARY

Two aspects of quality exist. One aspect is the quality of conformance to specification. The other aspect is the perceived grade of the product. Both impact the profitability of a company. Increasing the quality of conformance generally decreases the cost of making a product or providing a service. Investments made in preventing errors from occuring reduce the amount of scrap and rework.

A higher perceived grade of product, on the other hand, can impact profit by generating a price premium.

About 80% of errors occur because one of three conditions is not met. They are (1) knowing what to do; (2) having feedback on what was done; and (3) being able to modify actions if the feedback does not match what should be happening.

These three conditions are the criteria for self-control.

3

Beliefs About Improvements in Quality

Like most things, quality improvements don't just happen. They must be started by some one person and expand from there. Much has been written about the importance of top management in the implementation of a quality improvement program. A model of this method of quality improvement is shown in Figure 3.1.

Top management is the starting point. Management starts by establishing a policy statement that sets the direction for the company. This statement clearly states what the president expects from the employees and the final product. In the quality field it has been generally recommended that the policy statement be short and easy to remember. The policy statement for IBM, for example, states, "We will deliver defect-free, competitive products and services on time to our customers."

The objective of such statements is to let those inside the company

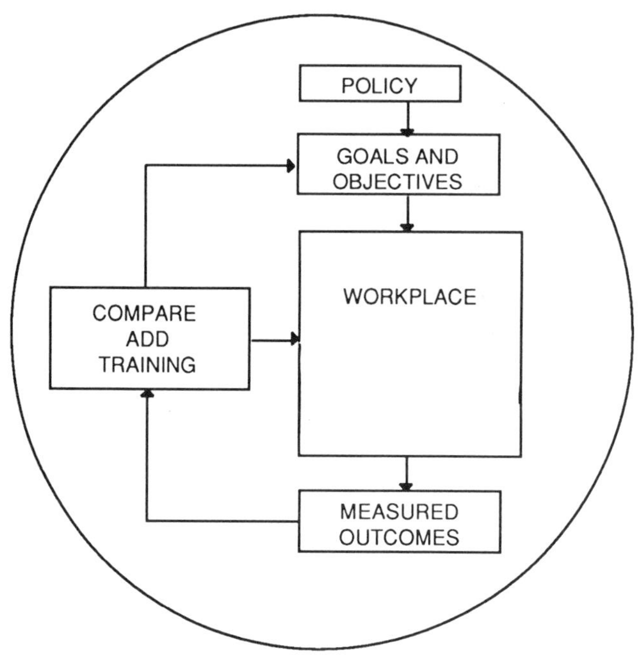

Figure 3.1 The quality improvement cycle.

know what is expected of them and those outside the company know what to expect from the company's managers.

Some of the advantages of having a written policy are

1. It forces top management to formalize its beliefs and desires about quality.
2. It provides a set of expectations from an authoritative source.
3. It establishes a framework that can be audited.

The last item is especially important, since the conclusions that everyone will form are dependent on deeds rather than words.

The next step in the product improvement cycle is to establish specific, attainable goals. These goals are defined and quantified to serve as a basis for planning. The advantages of planning are that it helps to unify the thinking of the managers, helps to allocate resources to meet the objectives, gives legitimacy to the work to be done, stimulates action on the objectives, and permits subsequent comparison of performance against objectives.

These objectives are communicated into the workplace. They become more specific as they are communicated down through the organization. Communication in the workplace tends to follow the organizational hierarchy within the company. A typical structure is shown in Figure 3.2. The starting point is the company president. The objectives flow down to the functional managers reporting to the president. This might include a Marketing manager, a Research and Development manager, a Manufacturing manager, a Finance manager, a Personnel manager, and a Quality manager as shown in Figure 3.2. These managers in turn establish and communicate goals and objectives within their organizations. At each step downwards, the objectives and goals be-

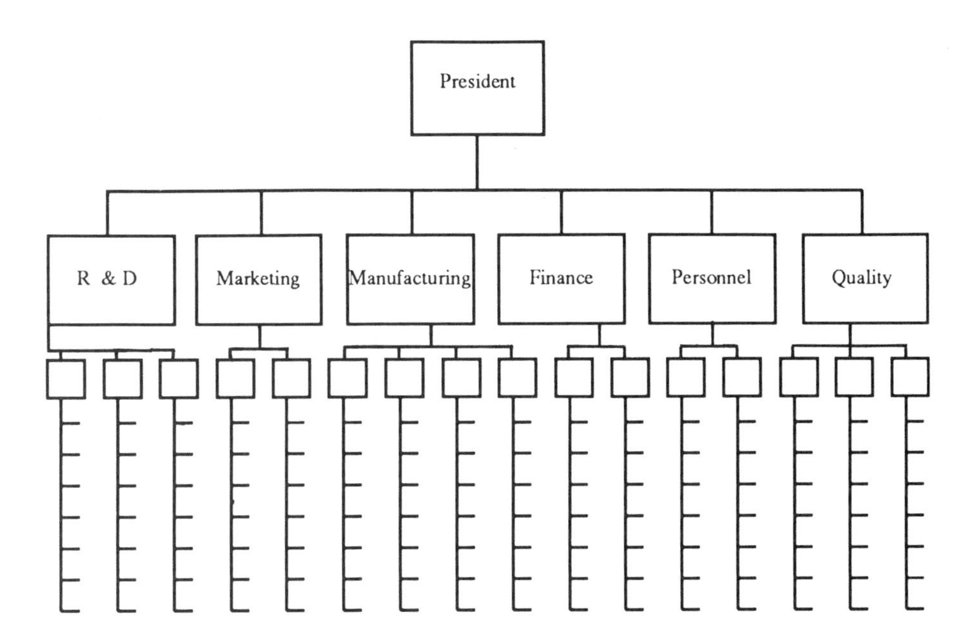

Figure 3.2 Communication flow within the workplace.

come more specific. For each objective or goal a measure is established to monitor progress.

As time passes, results are measured and compared with the objectives and measurements. This allows recognition to be given for accomplishments. It also helps to establish what the future objectives should be and what additional skills, resources, or training are needed to help achieve the objectives.

A COMPLEMENTARY QUALITY IMPLEMENTATION STRUCTURE

Complementing the organizational hierarchy is another structure for the implementation of quality. In this structure, you are the key. This structure is part of the network of interdependent partnerships that span the product flow. The smallest element in this working process I call the Quality Triad. It is shown in Figure 3.3.

At the center is the individual contributor—you. Immediately surrounding you is your process—how you do your job. Also working with you are two key people or sets of people. One is the customer; the other is the supplier. Each is surrounded by various processes.

Interactions occur as the product flows from your supplier to you

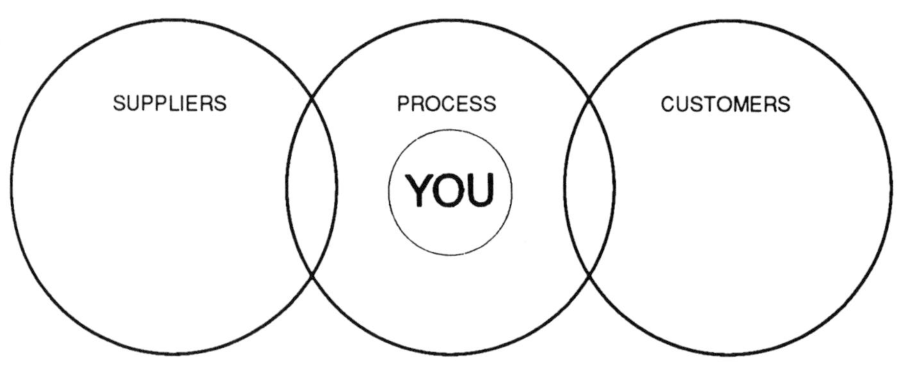

Figure 3.3 The Quality Triad.

and then to your customer. There is a blending of the supplier's process with your process and a similar blending between your process and your customer's process. There are also expectations and responsibilities that develop in these working partnerships.

Within this framework you play three different roles. You are a craftsman, a connoisseur, and an entrepreneur. The role of the craftsman is the role you play when developing a mastery of your trade or specialty. The craftsman's role emphasizes the relationship between you and your processes.

You assume the role of the connoisseur whenever you appreciate the subtleties of the products or services that you buy. This role emphasizes the relationship between you and your suppliers.

The final role, that of an entrepreneur, is the role you play when innovating and improving products with and for your customers. These roles provide the underlying structure for all quality programs. For it is the people at all levels operating in these roles that let the company goals and objectives happen.

SUMMARY

There are two complementary aspects to a quality program. One approach starts with top management and is driven through the organizational hierarchy by policy statements, goals, and objectives.

A second approach focuses on the smallest structure within the workplace, the Quality Triad. It starts with each individual contributor, a quality master. It expands to include those directly touched by the quality master and his or her processes, customers, and suppliers. This aspect concentrates on quality improvement through expectations and responsibilities for transactions in the product workflow. Each transaction directly affects quality.

4

Exploring the World of the Craftsman

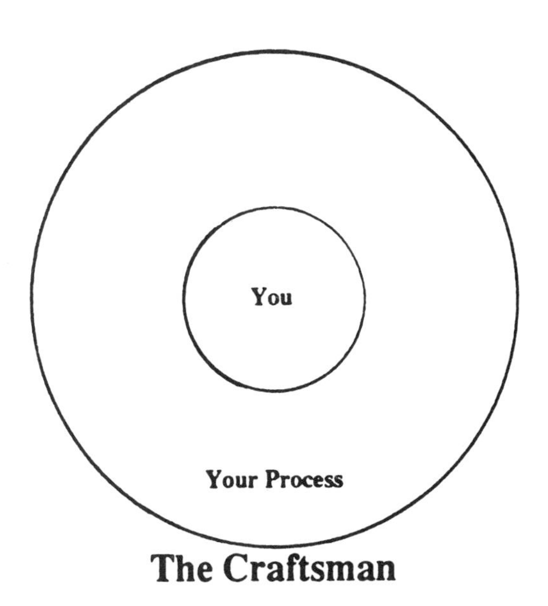

Craftsmanship is an old idea that still endures today. It means a special relationship between the person who works and the work done. Strictly speaking, a craftsman is a skilled worker in a craft or handicraft. In current usages, however, the term is often applied to any worker who excels in some art or profession. It reflects more the relationship between the person and the quality of the work than the relationship between the person and the type of work. With a craftsman there is an exercise of skill, ingenuity, and taste. Words like *handcrafted* reflect the care that such a person shows in the work.

I delight in watching someone work who is really good at the job. I enjoy seeing a beautiful product created or a service well performed. For me there is a certain magnetism in watching anyone who has turned work into an art. Such people show patience, care, and attentiveness, but there is also a oneness, a harmony, or a unity between them and their work. The work of an Olympic gymnast and an ice skater seems effortless, but the same potential exists in any job, from painting a house to painting a picture, from assembling a computer to designing a computer, from driving a diesel truck to driving a race car.

CHARACTERISTIC 1: THERE IS A HARMONY BETWEEN THE CRAFTSMAN AND THE WORK

The difference between a novice worker and a craftsman can be seen in the expression on the craftsman's face. One summer I worked on a steel gang laying track in Montana for the Northern Pacific Railroad. I was young and strong, and I thought that these traits would enable me to do a good job.

My first job was to drive spikes to secure the railroad track to the railroad ties that supported the track. I was one of two people who were to do this. The other man was Mossa, a sixty-two-year-old Italian who had immigrated to the United States and had worked since his early twenties on the railroad gangs. We each selected a spike maul, which is a specially designed sledgehammer for spiking. Mossa asked whether I had spiked before, and since I had not, he showed me how it was done. He tapped the spike through the tie plate and into the predrilled hole of the tie. When he felt the spike was tight enough he raised the spike maul over his shoulder and effortlessly came down directly on the spike. In six graceful strokes, the spike was pressing the rail. I watched him spike several in the same manner before I decided to try it myself.

I tapped my first spike to help hold it. I then gracefully raised the spike maul over my head and forcefully brought the spike maul down. Much to my surprise the spike maul rebounded seemingly faster than it went down. I soon realized that the spike maul had hit the rail rather than the spike. A bit more cautiously and with a little more concern on my face, I raised the spike maul again. This time I avoided the rail but hit the metal tie plate. Once again the spike maul flew back at me. I was beginning to get an idea of the skill required for this task. I was trying to hit a spike whose head was about an inch in diameter, when it was totally surrounded by (non-impact-absorbing) metal, with a spike maul whose head was about one-and-one-half inches in diameter. The expression on my face was as different from Mossa's as night and day. By the end of the summer I did get a lot better, but I still enjoyed watching the effortless spiking of Mossa.

CHARACTERISTIC 2: THERE IS A SELF-INVOLVEMENT WITH THE WORK

A second characteristic of the craftsman that I observed in Mossa was a self-involvement in what he was doing. There was no separation between him and the work. The motion of the spike maul and Mossa's thoughts were as one as he raised the spike maul up and glided it down onto the spike.

This self-involvement extends to all other jobs as well. It is the self-involvement that the engineer has when he or she unknowingly works through the lunch period; it is the concentration that a basketball player has in blocking out all outside distractions when attempting the final shot that will win the game. For a mechanic it is a sensitivity to the range of tightness for a threaded screw, from finger-tight to snug to tight, and an appreciation of not only the elasticity of metal but also its softness.

CHARACTERISTIC 3: THE CRAFTSMAN BELIEVES THAT THERE IS AN UGLY WAY AND A BEAUTIFUL WAY OF DOING ALMOST ANYTHING

The craftsman also believes that there is an ugly and a beautiful way of doing almost anything. Craftsmanship is more a journey than a destination. The craftsman believes in continual improvement. Craftsmanship is not just something he or she was born with, although its potential is inborn. It is also the result of listening and observing. It comes from directly seeing what looks good. It comes from exploring the underlying methods to arrive at what looks good. It comes from seeking why it looks good.

SUMMARY

A craftsman is any person who creates or performs with skill in his or her art or profession. Characteristics displayed by a craftsman are (1) a harmony between the craftsman and the work, (2) a self-involvement with the work, and (3) a belief that there is an ugly and a beautiful way to do any job.

5

Examining the Role of the Connoisseur

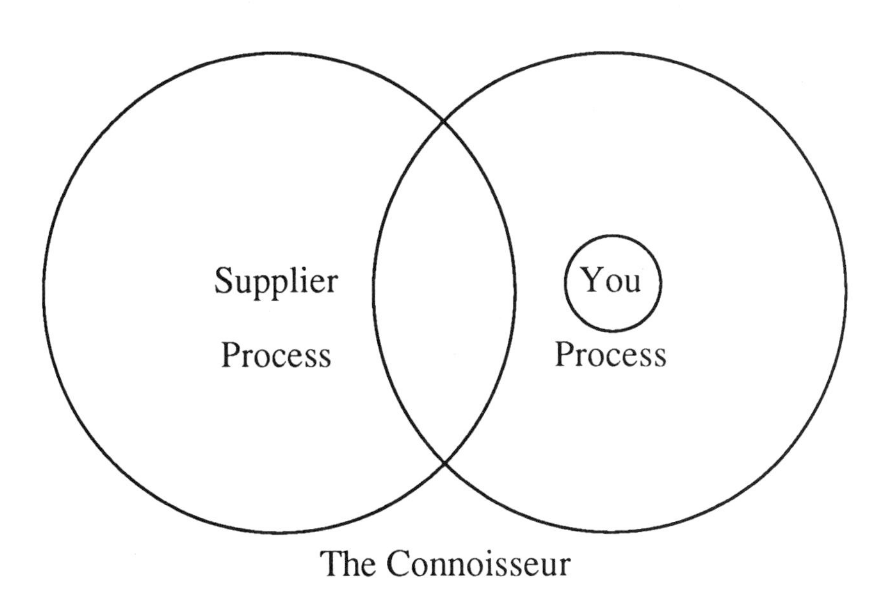

The connoisseur is one who enjoys with appreciation and discrimination the subtleties of the products and services he or she purchases. The connoisseur possesses a discerning eye for the best and a consuming desire for it. The connoisseur has learned to differentiate not only between bad and good but also between better and best. To achieve this, the connoisseur forges a partnership with his or her suppliers.

CHARACTERISTIC 1: THE CONNOISSEUR FORGES A CONSTRUCTIVE, HARMONIOUS, AND MUTUALLY BENEFICIAL WORKING RELATIONSHIP WITH SUPPLIERS

The connoisseur forges with suppliers a partnership that is constructive, harmonious, and mutually beneficial. Such a partnership is based upon equality and mutual dependence. It is not a relationship that pits power against power depending upon whether there is a buyer's market or a seller's market, nor is it a relationship in which one side has only obligations and the other only rights or entitlements. It is a relationship in which there are only obligations and all obligations have mutual obligations. Each side is obligated to provide what the other side needs to achieve its goals and fulfill itself.

These aspects can also be seen in the set of principles that the Japanese Union of Scientists and Engineers has developed to help build this relationship (Table 5.1).

CHARACTERISTIC 2: THE CONNOISSEUR RECOGNIZES THE IMPORTANCE OF THE SUPPLIERS

Yogi Berra, who at one time was the manager of the New York Yankees, was asked what it takes to be a good manager. His response: "Good players." This connection between being good and having good players on your team is generally acknowledged in sports. It is hard to find recognition for your achievements if your team does not do well. Baseball is an example. Toward the end of each season, a most valuable player for each league is named. In the last 10 years, half the winners of that award have been with a team that has either won the World Series or played in it. Only one player in twenty played for a team that finished lower than second in their division, and none of the winners of this award (until André Dawson of the Chicago Cubs) played for a team with a losing record.

This relationship also carries over to the workplace. Suppliers are an extension of your process and may be thought of as process partners. They help the work you do and can substantially affect your image as a craftsman. Developing this relationship allows you to "stand on the shoulders of giants" (the suppliers).

One of the keys to the improvement of the quality of Japanese products was paying attention to suppliers. Kaoru Ishikawa, one of the

Table 5.1 Ten Quality-Control Principles For Vendee-Vendor Relations

---

Preface: Both vendee and vendor should have mutual confidence, cooperation, and the high resolve to live-and-let-live based on the responsibilities of enterprises for the public. In this spirit, both parties should sincerely practice the following "Ten Principles."

Principle 1: Both vendee and vendor are fully responsible for quality control applications with mutual understanding and cooperation between their quality control systems.

Principle 2: Both vendee and vendor should be independent of each other and esteem the independence of the other party.

Principle 3: Vendee is responsible to bring clear and adequate information and requirements to the vendor so that the vendor can know precisely what he should manufacture.

Principle 4: Both vendee and vendor, before entering into business transactions, should conclude a rational contract between them in respect to quality, quantity, price, delivery terms, and method of payment.

Principle 5: Vendor is responsible for the assurance of quality that will give satisfaction to vendee, and he is also responsible for submitting necessary and actual data upon the vendee's request.

Request 6: Both vendee and vendor should decide the evaluation method of various items beforehand, which will be admitted as satisfactory to both parties.

Principle 7: Both vendee and vendor should establish in their contract the systems and procedures through which they can reach amicable settlement of disputes whenever any problems occur.

Principle 8: Both vendee and vendor, taking into consideration the other party's standing, should exchange information necessary to carry out better quality control.

Principle 9: Both vendee and vendor should always perform control business activities sufficiently, such as an ordering, production and inventory planning, clerical work and systems, so that their relationship is maintained upon an amicable and satisfactory basis.

Continued

Table 5.1 *Continued*

Principle 10: Both vendee and vendor, when dealing with business transactions, should always take full account of consumer's interest.

Source: Ishikawa, K. *What is Total Quality Control the Japanese Way?* Englewood Cliffs, N.J.: Prentice-Hall, 1985.

foremost authorities on quality in Japan, put it this way: "The genesis of Japanese high quality was the selection of suppliers who had a good quality program."*

One reason for this is that suppliers provide a major portion of the total product. In Japan, for example, the proportion of the total product cost that a company buys from outside suppliers is about 70%. In the United States and Europe, the proportion is not quite as large; it is about 50%, but this is still significant. The same is true in the Quality Triad (see Chap. 3). A major portion of what we accomplish is based on the contributions of others.

IDENTIFYING YOUR SUPPLIERS

People can become suppliers in several ways. The most common is to supply resources that we are most likely to use, such as machines or equipment, materials or information, methods or procedures, manpower, or a certain environment. Sometimes these suppliers expand what we really like to do. In writing this book, for example, I have relied on suppliers of information. In a sense, I have been able to stand on the shoulders of giants. People like Juran, Ishikawa, Deming, Crosby, Peters, and Tice have not only helped my efforts but also provided role models.

* Ishikawa, K. *What is Total Quality Control the Japanese Way?* Englewood Cliffs, N.J.: Prentice-Hall, 1985, p. 156.

Other times people can contribute by supplying services for things that I do not like to do or things that I do not do well. One example is the way in which I used to maintain my car. For years I tried to maintain my car myself, but it was always a task that got neglected. During this time it was fair to say that neither the reputation of my car for reliability nor my own reputation as a craftsman mechanic was enhanced.

Then I changed my approach to car maintenance. I realized that somewhere out there was someone who loves taking care of cars. I found Vic, and we have developed a long-term car-care relationship. Since then, my reputation for supplying dependable transportation has improved significantly.

Buying a service is just one of the ways that we can stand on the shoulders of giants. We can also improve our efficiency through the tools of machines that we use. I used to maintain yards on summer jobs when I was in high school. I would mow a lawn and then trim its edges. Mowing the lawn was fairly easy since I had a power lawn mower, but the trimming was a long, taxing ordeal using the hand clippers. Years later, when nylon string trimmers such as Weedeaters were introduced, I was one of the first to buy one and thereby stand on the shoulders of giants and improve my efficiency.

Some tools are physical tools like the weedeater, but others are more intellectual, as for instance a calculator instead of a slide rule or a text editor instead of a typewriter. Still others are conceptual tools or new methodologies like statistical quality control or just-in-time delivery. And the leverage that these tools provide compared to older ones is like night and day.

CHOOSING A SUPPLIER

In choosing a supplier, there is a combination of good news and bad news. The bad news is that in many situations you may not be able to choose a supplier at all. The good news, however, is that very often, as in marriage, it is not so much a question of choosing the right partner as of being the right partner.

In this sense, the family has much in common with industry. Partners are not selected on the basis of any psychological testing. There is little latitude for alternatives in people that can be chosen, yet through the proper set of expectations and responsibilities, high-quality work can still result.

One example stands out in my mind. Alisa, one of my daughters, decided that she would like to earn some money. She saw that the house needed to be vacuumed and knew that she could do a good job, so she offered to vacuum the entire house. Her mother, Diane, agreed to let her earn money this way. Alisa was to vacuum the house completely, moving furniture and even going along the walls with a special attachment. After Alisa was done, she was to get Diane to check it. Alisa proceeded to work. She was doing quite well until the phone rang. It was Cheryl, Alisa's best friend, who wanted Alisa to go shopping with her.

This event transformed an entrepreneurial craftsman into a questionable supplier. After a few quick swooshes with the vacuum cleaner here and there, she decided she was done and went to get her money. It was easy to see that this was not the job agreed to earlier. "Wait," Diane said. "This is not the type of work you do. There's an ugly and a beautiful way of doing everything. I expect a beautiful job—and an assurance that it will be right." Although Alisa did not say a word, you could read the words on her face: "Obviously you've confused me with someone who really cares." Undaunted by this, Diane knew that the caring supplier in Alisa would eventually appear. Rejecting carelessness and substituting the proper image of what needed to be done, she forged the beginning of a partnership. That incident occurred five years ago. Since that constructive time Alisa has become a craftsman, meticulous in her work and dependable for all tasks she does.

THE CONNOISSEUR BELIEVES THAT THERE IS NO SUCH THING AS A COMMODITY

The connoisseur realizes that when something is considered as a commodity, as something that requires no special attention, it will become

Isn't it amazing what can go wrong with a simple screw?

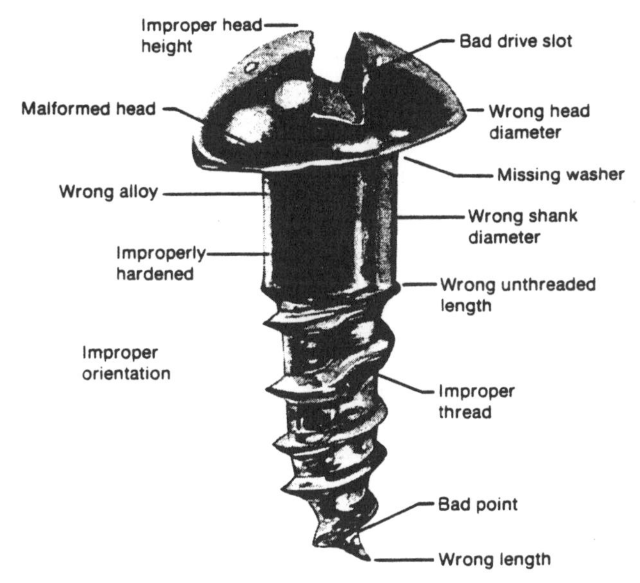

Figure 5.1 (Source: Magnetic Analysis Corp.)

exactly that—nothing special. I fell into this trap. I needed a couple of wood screws to secure a storage shelf to the wall in the garage, so I went to a small convenience store to pick up some of simple wood screws.

What I found out was that all screws are not created equal. The screws that I got turned out to be made of a soft metal that let the head deform as I tried to screw them into the wall. This is just one way that the simple generic screw is neither generic nor simple (Fig. 5.1). The moral is: Don't let simple/inexpensive parts, improperly made, interfere with production.

SUMMARY

A connoisseur is one who enjoys with appreciation and discrimination the subtleties of the products or services he or she purchases. The connoisseur believes in forging a constructive, harmonious, and mutually beneficial relationship with suppliers. The connoisseur recognizes the importance of suppliers so that he or she can improve their work.

Suppliers are people who provide resources to do the work: machines or equipment, materials or information, methods or procedures, or the working environment. Connoisseurs believe there is no such thing as a commodity.

6

Examining the Role of the Entrepreneur

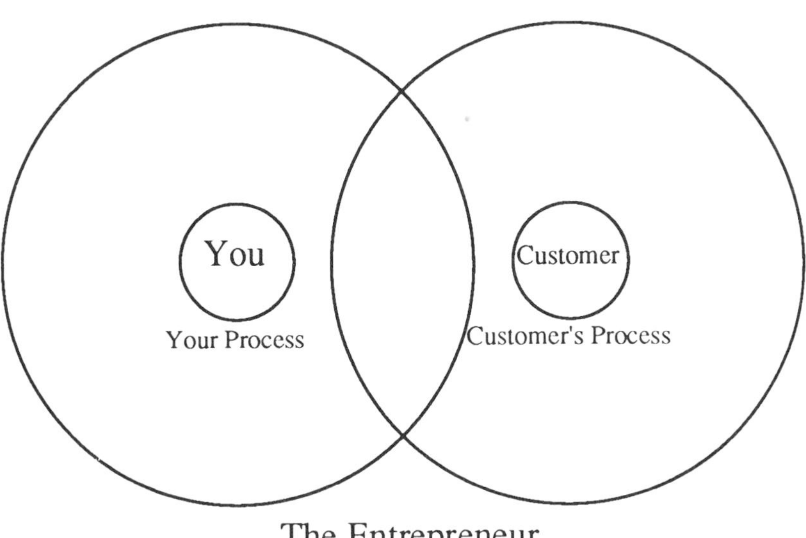

The Entrepreneur

THE CONTRIBUTIONS OF THE ENTREPRENEUR

There is an old story of three stonecutters who were asked what they were doing. The first replied, "I'm making a living." The second kept on hammering while he said, "I'm doing the best job of stonecutting in the entire country." The third one looked up with a visionary gleam in his eyes and said, "I'm building a cathedral." This third person is also the third person in our Quality Triad—the entrepreneur.

THERE IS A VISION FOR THE CONTRIBUTION PROVIDED

From the steelworker in Pittsburgh to the launch team member for the NASA shuttle in Florida; from the textile worker in North Carolina to the wheat farmer in North Dakota; from the electronics technician in San Jose to the auto assembler in Detroit, each worker makes a contribution out of unique talents and circumstances. Each recognizes his or her own contribution. The steelworker in Pittsburgh, for example, is not just melting ore but also providing the building blocks for items ranging from buildings to barges, from trains to tanks, from cars to cables.

THE ENTREPRENEUR REALIZES THAT EVERY TRANSACTION FORGES A PARTNERSHIP

Like the connoisseur, the entrepreneur believes that the relationship between the buyer and the seller should be mutually beneficial. The entrepreneur also realizes that each transaction helps either to forge a partnership or to weaken the partnership.

I once attended a workshop on how to satisfy customers. We were asked to describe an experience that stood out because of the kind of service, either positive or negative, that was received. Such an exercise is one of the best ways to discover places to go for good service, and it also showed the power of a single transaction. One might later get a large account because of a small act of courtesy. If a return is handled courteously, a disappointment can turn into confidence in the entrepreneur. We should concentrate not on the product but on forging bonds of trust and loyalty. The entrepreneur realizes that in any business, from selling cars to selling major defense systems, 80% of the transactions are with repeat customers. But this is not the sole reason that the entrepreneur values the wants of the customer.

THE ENTREPRENEUR BELIEVES THAT THE NEEDS OF THE CUSTOMER ARE PREEMINENT

The entrepreneur is building cathedrals, not polishing stones. If, in the eyes of the person who is to use it, the task is the wrong one or not useful, the entrepreneur responds. This can be seen in the final principle of the J.U.S.E. list (see Table 5.1). Both vendee and vendor, when dealing with business transactions, should always take full account of the consumer's interest. This reflects the need to satisfy the customer with the product or service. If a change has to be made, the customer can only try to correct an incorrect product or service. The fundamental changes that must be done to make the customer satisfied the first time cannot be done by the customer. They must occur at the process level. If the customer ends up doing the fixes, the improvement process cannot occur.

IDENTIFYING YOUR CUSTOMERS

In the open marketplace, the exchange of money for services or products defines who is the customer and who is the supplier. Within a company, this means of identifying the customer becomes harder because there is not necessarily an exchange of money involved.

One alternative method is to identify as a customer that person who relies upon you to accomplish their job. One obvious person who would fit into this category is the next person in the process following yours. In the manufacturing environment, this would be at the next step in the manufacturing flow diagram. For staff members in such departments as Personnel, Accounting, Production, Engineering, and Quality Control, there are two sets of customers. One set is the president or factory managers to whom they submit plans and proposals. The other set is the first-line organizations such as the Design, Purchasing, Manufacturing, and Marketing departments that they support.

SUMMARY

The entrepreneur creates customers and builds customer loyalty. The entrepreneur has a vision of the services or products to be provided. The entrepreneur realizes that every transaction forges a partnership. The entrepreneur believes that the needs of the customer are preeminent. The customers are the people who use the products or services.

7

Linking the Quality Triad with the Quality System

Two aspects of quality have been looked at. One started with top management and is directed down through the organizational hierarchy. The other starts in the workplace and follows the product flow. These two aspects are pictured together in Figure 7.1. They are pictured together because they complement one another. The outward structure connects the individuals with the overall mission of the organization. The underlying structure, the Quality Triad, reflects the interdependent working structure that supports the company's mission.

The complementary aspects of these two views are also shown. The quality system directs itself at product issues. It deals with things. It starts with objectives—quantifiable, unemotional, and measurable things. It focuses on the desired end results. It addresses the science of work.

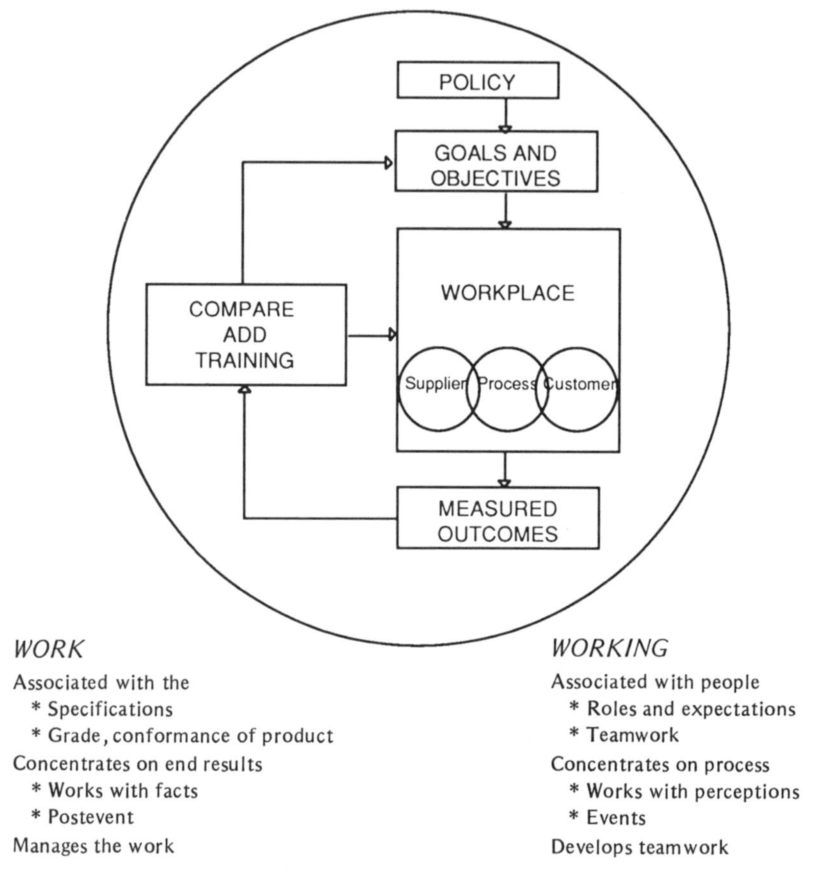

Figure 7.1 Linking work and working.

The Quality Triad complements the science of work by looking at the art of working. It complements the product issues to include issues related to people. It complements the issues of grade and conformance of products to include issues related to the roles people play and the expectations and responsibilities of those roles. It complements the post-event product quality to the real-time activities associated with producing that product. It complements the financial benefits associated with product quality to the sense of achievement enjoyed by the craftsman, connoisseur, or entrepreneur.

LINKING THE DIFFERENT ASPECTS OF QUALITY TOGETHER

The complementary aspect of these views can also be seen in the linkages that connect the financial measures of quality with the roles in the Quality Triad. The systems approach started with one of the requirements for a business—profitability. For this, quality has two major impacts. Two complementary dimensions were explored. One was the effect on income, the other the effect on the cost to produce the product. The key ideas associated with the financial impact are summarized in the financial grid shown in Figure 7.2.

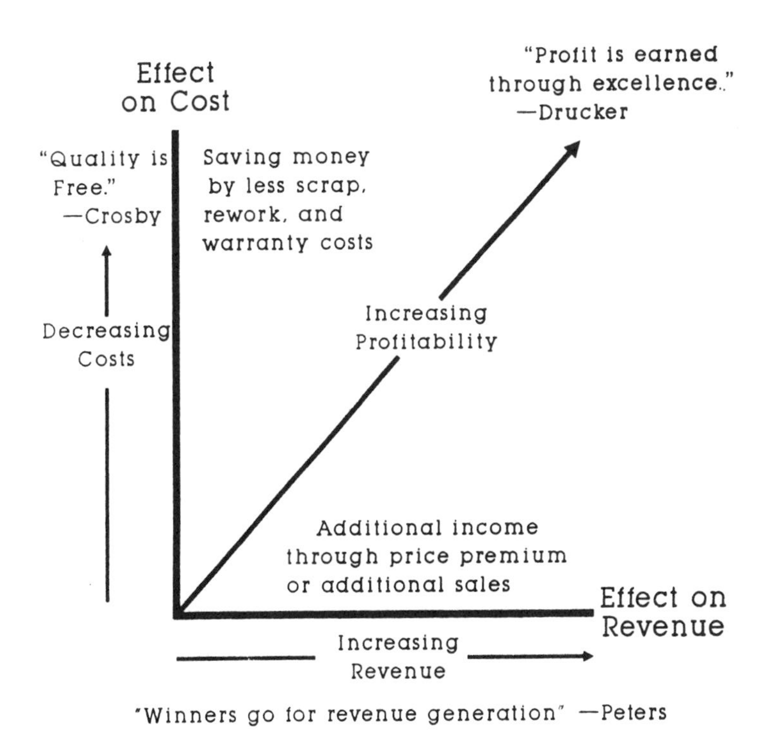

Figure 7.2 The financial impact of quality.

These two dimensions translate into two dimensions of product quality. The effect on income relates to the grade of quality as perceived in the marketplace. The cost of quality relates to the conformance of the product to the requirements. The relationship between these two aspects of quality is shown in Figure 7.3.

Three key factors determine product excellence. These factors are

1. The processes used to produce the product/service.
2. The customers who use the product/service.
3. The suppliers who become part of the process.

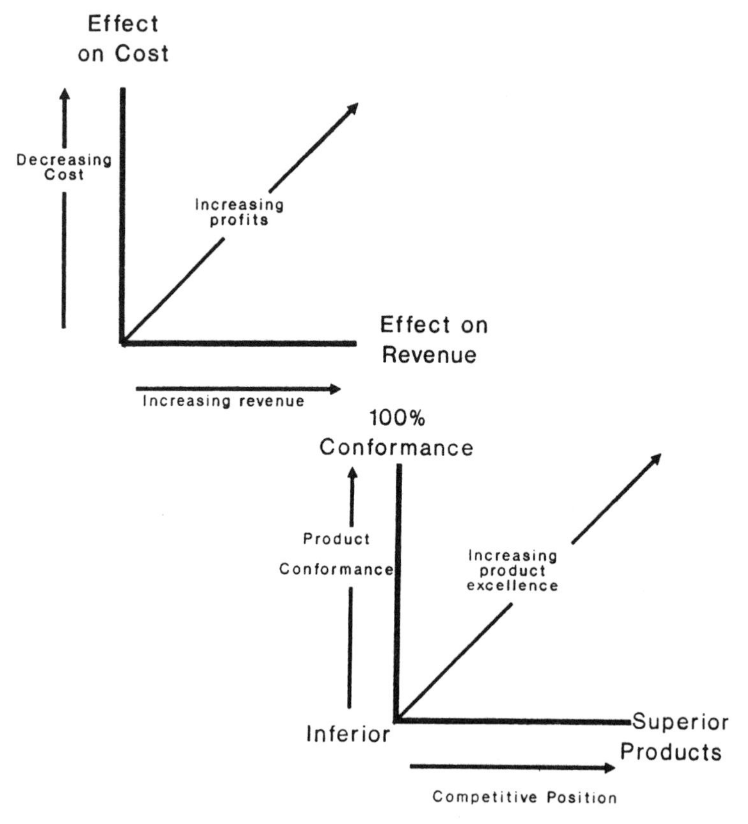

Figure 7.3 Linking the financial impact with product excellence.

Three key factors are shown in Figure 7.4.

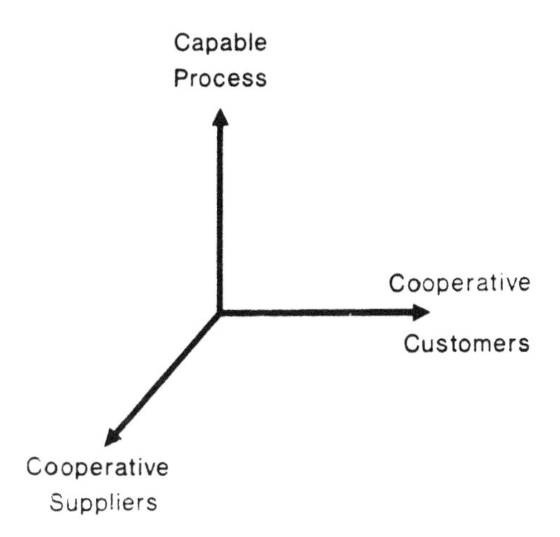

Figure 7.4 Three factors establishing product quality.

The next linkage connects the product characteristics with key factors for product quality. The grade of the product in the marketplace is the perception of the *customers.* The conformance of the product is the result of the *process* used to produce the product. The process in turn can also be divided into processes done within the factory and those done outside the factory by *suppliers.* The linkage between these dimensions and the product grid are shown in Figure 7.5.

The next linkage connects the roles in the working relationships between customer, process, and supplier. The role of creating a loyal, satisfied customer is played by the entrepreneur. The role associated with a stable, capable process is played by the craftsman. Finally, the role of a responsive, stable supplier is played by the connoisseur. These linkages are shown in Figure 7.6.

The final linkage is the recognition that at the center of these roles is you, the individual contributor. This is shown in the Quality Triad. It is you who is the craftsman with your process, the entrepreneur with your customers, and the connoisseur with your suppliers. See Fig. 7.6.

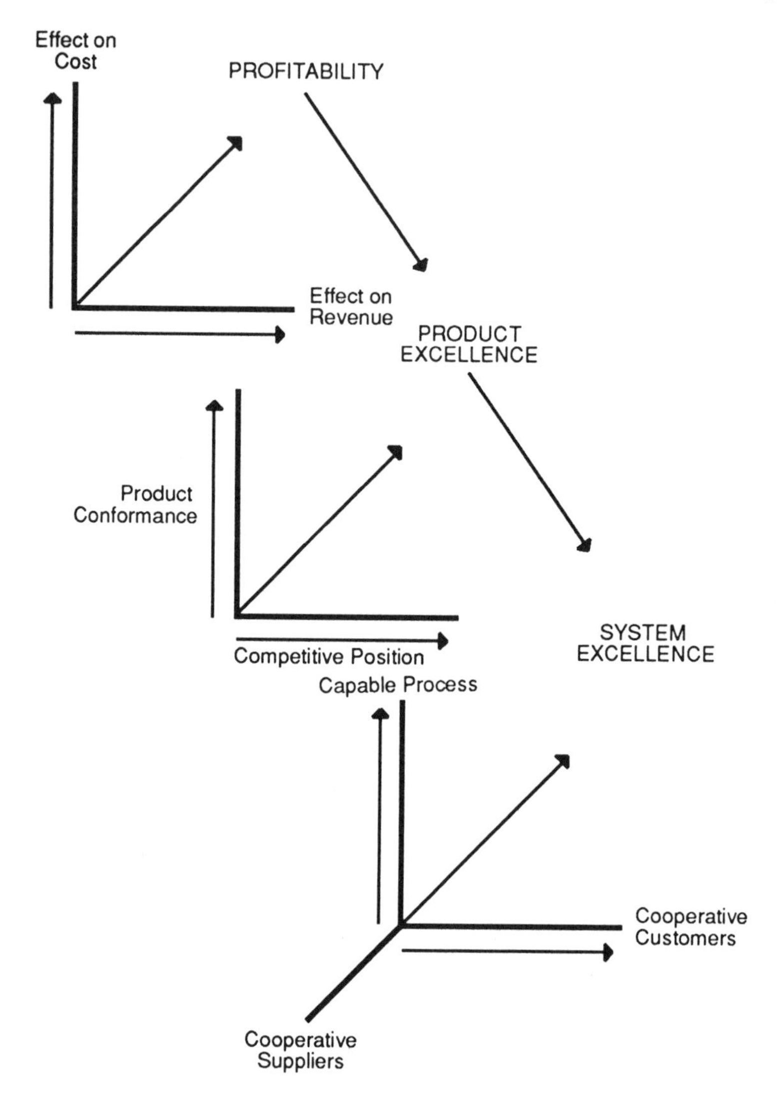

Figure 7.5 Quality linkages.

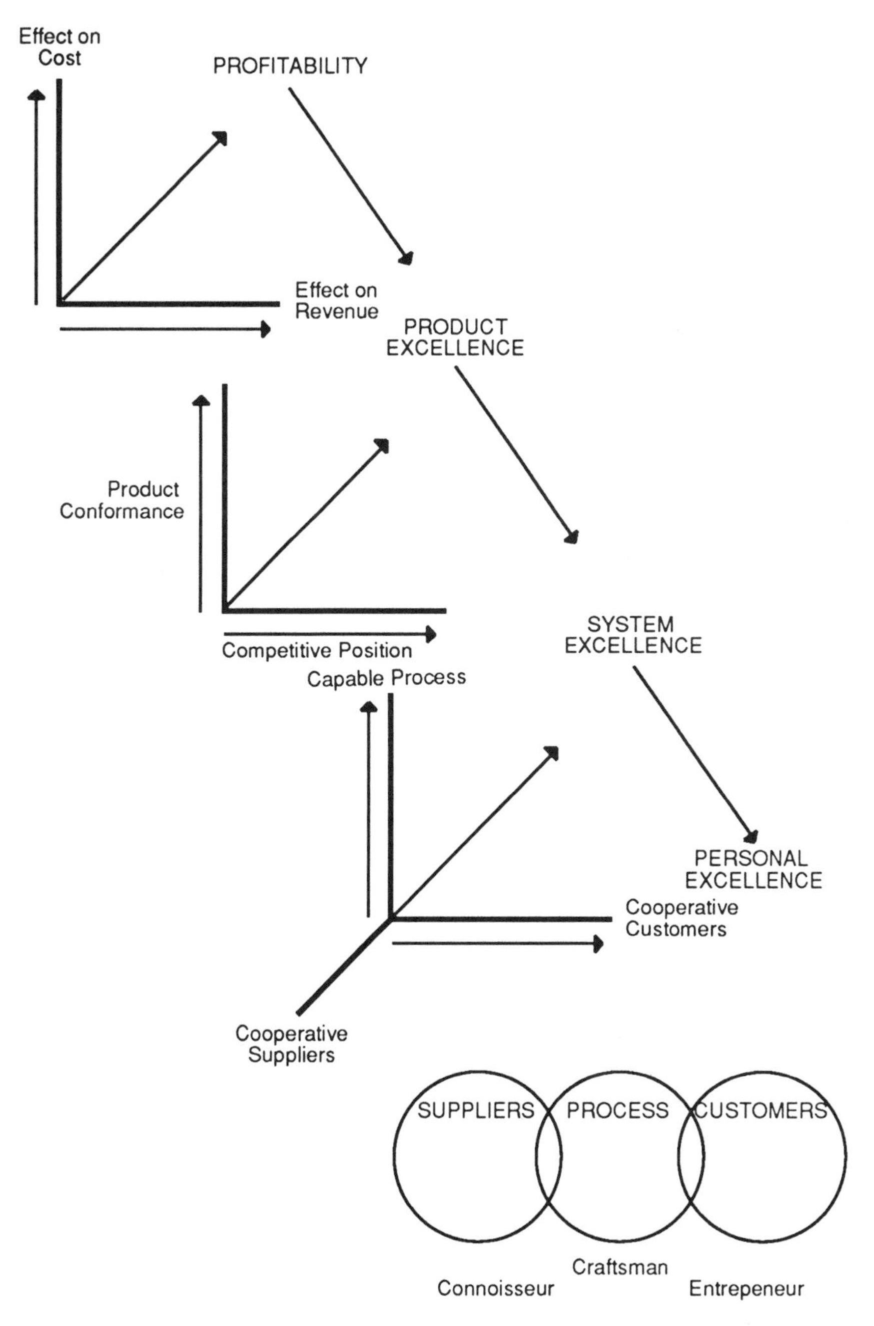

Figure 7.6 Linking the Quality Triad with the Quality System.

SUMMARY

The Quality Triad and the quality system are complementary aspects for making improvements. The quality system is directed at product issues. The Quality Triad focuses on people, their roles and their expectations.

The complementary aspect of these two approaches can be seen by a series of linkages. The financial impact of quality is linked to product excellence. One step removed from product excellence are the key factors determining product excellence: a capable process, cooperative suppliers, and satisfied customers. Linked to these three key factors are the role models comprising the quality master: the craftsman, the connoisseur, and the entrepreneur.

Part II

EXPECTATIONS

8

Expecting the Best

EXPECTATIONS

Closely related to beliefs are expectations. Expectations are regulating mechanisms for our behavior and have a very powerful influence on what we accomplish. Expectations serve as preset control limits. They can regulate our behavior as a thermostat regulates the temperature in a house. A diagram of how the temperature is regulated is shown in Figure 8.1.

The thermostat is preset to the desired temperature. A heat sensor measures the temperature in the house and feeds this information to the thermostat, which then compares the actual temperature with the preset temperature. If the temperature in the house is more than the minimum temperature and less than the maximum temperature, no action is taken. If the temperature is below the preset temperature, the heater is turned on. If it is more than the preset limit, the heater is turned off.

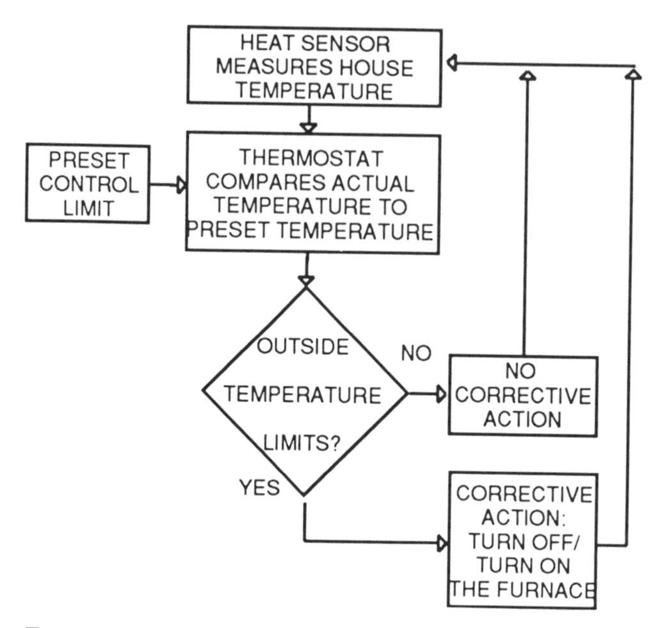

Figure 8.1 How control limits regulate house temperature.

HOW EXPECTATIONS AFFECT OUR BEHAVIOR

The corresponding diagram for controlling our actions is shown in Figure 8.2. In this system, the preset limits are our expectations. They represent what we expect to happen. When we observe an event, our minds compare our perception of what happened with what was expected. If the difference between the two is small, no action is taken. If, on the other hand, the difference is large, we act to get them to agree.

AN EXAMPLE OF EXPECTATIONS REGULATING BEHAVIOR

Heidi, my five-year-old daughter, showed me how her expectations regulated her actions. I took her with me to the grocery store, and as we

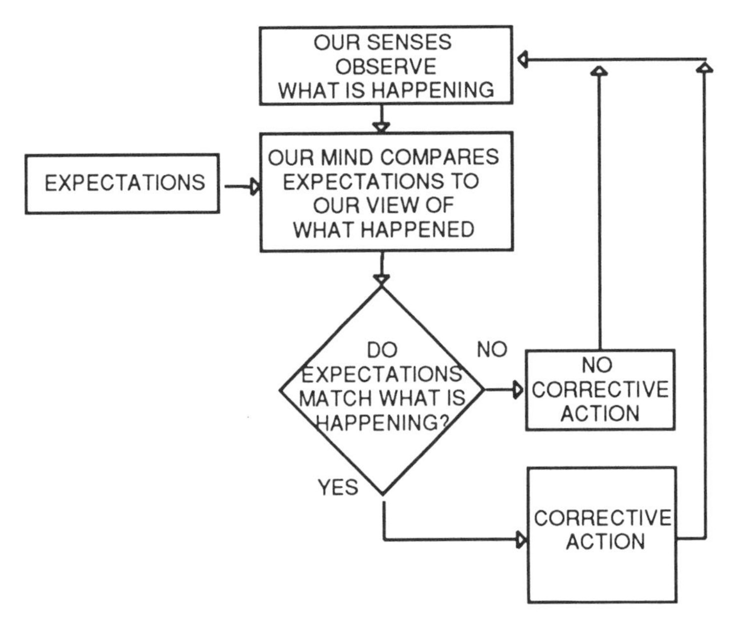

Figure 8.2 How expectations affect our behavior.

went down the cereal aisle, she proceeded to convince me that we were about to run out of her favorite cereal, the one with the best prize inside. Up and down the cereal aisle she went until she found it. Then all during the time we were in the store she admired the picture of the stickers that were enclosed. She anticipated the joy and fun she would have decorating her toys with these stickers. Finally we got home. Quickly she found the box of cereal and gave it to me to open. I opened the box and looked for the free prize, but it was not to be seen. Encouraged by Heidi that it must be there, I emptied the entire box of cereal into a large bowl. The stickers were still not to be seen. Carefully now I poured back the cereal into the box looking for the prize. Back and forth the cereal went from bowl to box while we were looking for the prize. Finally, Heidi took the box from me and proceeded to examine the box itself. She took out the wax paper wrapping within the box and found the stickers.

The goal had been achieved, and the regulating mechanism that got us there was Heidi's expectation that within the box there would be free stickers.

In this incident, Heidi had definite expectations that there should be stickers in the box. She was willing to pour the cereal back and forth several times to find the stickers. She was strongly motivated because her expectation was not being met. If, on the other hand, her expectation had been that sometimes the prize was accidentally omitted, she might not have found the stickers.

LOW EXPECTATIONS DECREASE ACHIEVEMENTS

Expectations that are set too low can inhibit the ability to achieve the best. Consider the baseball batter who expects to hit a yearly average of .275 but suddenly finds himself hitting .325. If the expectation of what he is able to hit is .275, suddenly the pressure is on. He will ask himself what is going on. He will be uncomfortable and most likely will become very creative at reducing his batting average to one within the set of expectations.

Life is full of such examples. Suppose I am asked to apply for a position for which I do not think that I am qualified, although others do think so. Chances are I will do all I can, subconsciously and inadvertently, to decrease my chances of getting that new position—and satisfy my expectations of what I should have.

EXPECT THE BEST

We can choose our own expectations or we can let them be formed by others. One choice is to expect the best. Expecting the best causes us to react whenever we do not get what is best for us.

Expecting the best creates a set of expectations that lets good things be drawn to us. Norman Vincent Peale states it this way: "When you expect the best, you release a magnetic force in your mind which by a law of attraction tends to bring the best to you.—Every great thing becomes for you a possibility."*

An alternative is to set low expectations. Then we will accept situations that do not allow us to be or do our best. In fact, we may react negatively to situations that are "too good" for us. It is in this way that by not expecting the best, good will not be drawn to us.

Expecting the best is also a good first rule when dealing with people. We can communicate either a positive or a negative expectation to people. In this connection we speak of the Pygmalion Effect or the self-fulfilling prophecy. George Bernard Shaw's play *Pygmalion*, which was later made as the basis for the musical comedy *My Fair Lady*, is one illustration of this. In the play, Professor Henry Higgins changes Eliza Doolittle from an uncultured flower girl into a polished lady through his belief in his own ability and his expectation of influencing Eliza's behavior.

SUMMARY

Expectations serve as preset control limits that regulate our behavior. When what we are observing differs substantially from our expectations, we take action to get them to agree.

Believing that you deserve the best and expecting the best create a magnetism that draws good toward you.

*Peale, N.V. *The Power of Positive Thinking*, New York: Prentice-Hall, 1956, p. 14.

9

Expectations in the Quality Triad

I have mentioned three roles in the Quality Triad. There is one more. This is the wizard. The role of the wizard is significant because the expectations of the wizard are different from those of the other roles. The wizard has set low expectations but tries to overcome them by believing in the magical power of transformation. Like the wizards in the days of the Knights of the Round Table, the wizard believes in his own magical power, the ability to transform a sow's ear (or any other starting material) into a silk purse. In fact, the wizard believes so much in the ability to transform an object that the wizard does not spend time preventing errors.

Now I know (in an intellectual sense) that I do not have this magical power within me, but sometimes I act as though it were there. I take my family out for lunch at McDonald's. I have done this before, so I know what to expect, and I am always ready for what follows.

McDonald's prefers to serve the Big Mac with two all-beef patties, special sauce, lettuce, cheese, pickle, and onions on a sesame seed bun. My children, however, do not quite see their Big Mac that way. They order it without the special sauce. This is a simple request, but it is one that requires special attention. Most of the time it is fulfilled, but not always. Then the wizard appears. With the stroke of a knife he removes the special sauce from the hamburger and provides a hamburger bun without sauce, thus transforming a Big Mac with special sauce into a Big Mac without special sauce. Now this may not rival some feats of magic that you have seen, but it is a transformation of a kind: a problem was corrected. I as wizard was able to transform an undesired product into the desired one.

This is not the only time that the wizard has the opportunity to appear. The wizard can appear whenever I rely solely on correcting a problem after its occurrence rather than working to prevent it before hand. For example, I recently went to buy a picnic table, because the picnic table I already had was too small, and it had gone through several repairs. In the first place I looked, I found a picnic table that was big enough, but it was held together with nails, and on the old table I had had to replace the nails with screws. The wizard within me, however, realized that once again I could rescue the situation. I could transform this nailed, soon-to-fall-apart table into one held together with screws that would last forever. An hour of work at the most, I thought (but I very often underestimate how long a job will take), and the transformation would be complete.

REVIEWING THE CONTRIBUTION OF THE WIZARD

Did the wizard satisfy everyone in the process? In both cases I eventually got what I wanted. The Big Mac was now without the sauce and the picnic table would be solid. McDonald's and the hardware store did sell their products. But we all lost something. I had to do extra work to make these transformations, and I will probably have to do the same sort of thing again. Consequently, the sellers probably think that they fulfilled my needs, when in fact they may lose my future business because of the problems I had.

Similar situations can also occur in the business environment. Businesses, however, do not believe in wizards. Businesses believe instead in certain special factories, but since these are not always seen, they are called hidden factories. These are factories within the standard factories, and they do all the rework—redoing the things that are not right the first time. There are several estimates of how large these hidden factories are. Phil Crosby estimates that 20% or more of sales in manufacturing companies and 35% of operating costs in service companies is spent in these hidden factories.*

Most of these studies have analyzed the manufacturing process. Manufacturing, however, is not the only place where these factories appear. I have seen them appear in services. I see them when I have a meeting and find that the room that was to have an overhead projector does not, or that the room that was reserved was double booked, or that the handouts that were to be prepared were not, and on and on.

SOURCES OF WASTE

The above are examples of supplier-induced error. These are also process-induced errors, and customer-induced errors (see Table 9.1).

Supplier-induced errors can be further subdivided into material-supplier-induced errors, environment-induced errors, machine-induced errors, and so on.

OTHER TYPES OF WASTE

Errors are not the only things that create waste. There are at least two other sources of waste: *erratic levels of work* and *complexity*. William

* Crosby, P.B. *Quality Without Tears: The Art of Hassle-Free Management*, New York: Plume Books, 1984, pp. 85–86.

Table 9.1 Sources and Types of Waste

	Types		
Sources	Errors	Complexity	Erratic levels of work
Supplier-induced	Supplier-induced	Supplier-induced	Supplier-induced
Process-induced	Process-induced	Process-induced	Process-induced
Customer-induced	Customer-induced	Customer-induced	Customer-induced

Conway, past Chief Executive Officer of Nashua corporation, has estimated that the productivity loss associated with each of these categories is 15%. That is, 10–15% of all work is associated with error, another 10–15% is associated with erratic levels of work, and yet another 10–15% is associated with complexity.*

UNEVEN LEVELS OF WORK AS SOURCES OF PRODUCTIVITY LOSSES

The productivity losses due to uneven levels of work are easily seen whenever you find yourself standing in a line and waiting, whether you wait in a checkout line in a grocery store, in a line at a busy ice-cream store, or in a factory at a copying machine.

Most of the time we do not document how much time this takes. When we do this, it can be surprising. In one of my classes, there was

* Conway, W., in "Nashua Seminar," Detroit: Ford Motor Co.,(1981), available from duplication dept.: #P1584.

one person who kept track of the number of ingots to be tested (work level) and how it varied with time. The graph is shown in Figure 9.1

The dramatic fluctuations of workload create low productivity in two ways, first, during periods when work is not available, waiting times occur; second, during periods of excessive work, additional space is needed to store work to be done, and systems have to be developed to control and identify the work to be done.

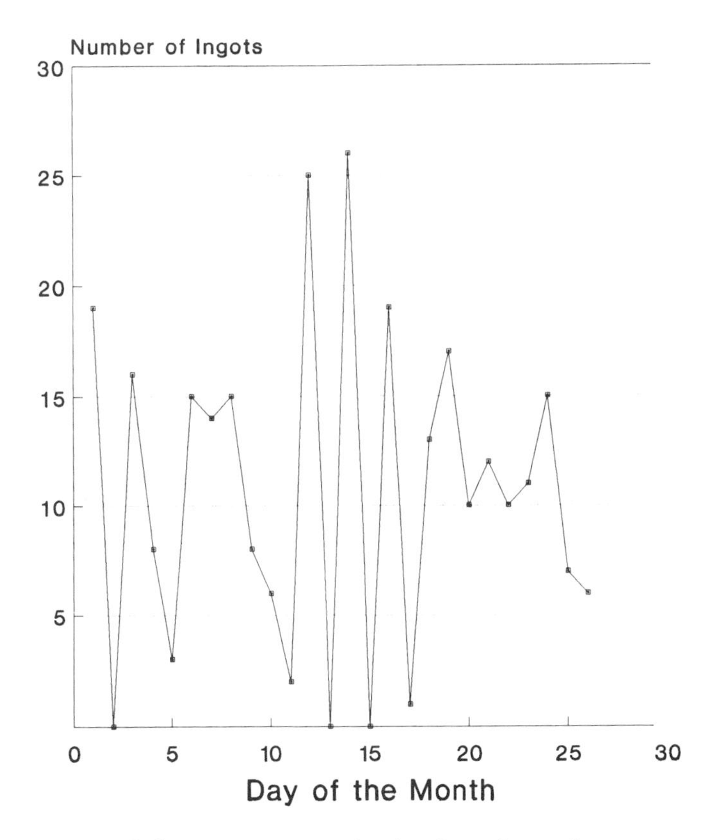

Figure 9.1 Variation in production from day to day.

COMPLEXITY AS A SOURCE OF PRODUCTIVITY LOSS

Complexity Due to Additional Information Needed

Most of us experience the issue of complexity at least once a year. As April 15th approaches, most people get to fill out at least one form in which there are no errors—but in which there can be a fair amount of complexity—the income-tax form. In this form it may take a page of small print even to determine whether you qualify as the head of a household. It is not just errors that take up resources and contribute nothing. Complexity has the same effect. The complexity introduced here results from the amount of information needed to fill out the form.

Bureaucratic forms are not the only things that perpetuate complexity. Industries also have their examples. The early (non-user-friendly) computers provide other examples of complexity. With these computers there are not any errors, but it may require three manuals, a service technician, and two hours to set up the computer to type the first letter.

Complexity Introduced by Variability

A second factor introducing complexity is the amount of variability associated with a process. The larger the variability, the greater the complexity. In this situation, productivity is decreased because the variabilty requires increased planning and uses additional resources. The matter could be as simple as not being able to predict the weather for a trip and so having to take a larger wardrobe or buy additional clothes during the trip. In an industrial setting, variability requires additional planning and setting aside of capacity if the yield from a process is erratic and unpredictable.

Complexity Introduced by Increasing Separation of Key Interfaces

A third factor introducing complexity is the lack of proximity of the key elements in the work in either time or distance. The farther away you are, the greater the complexity. For example, customers or suppliers who are within a few feet of you are easier to talk to than those who must be reached by telephone or visited miles away.

Likewise, processes or procedures that take a long time for the outcome to be measured or monitored are more difficult than ones that give immediate feedback. For example, a glue that sets in seconds is easier to monitor than one that must set for several hours.

Complexity Introduced by the Quantity of Key Relationships to Do the Work

A fourth factor introducing complexity is the quantity. The larger the number of different customers, suppliers, products, or processes, the greater the complexity. For example, if there are four suppliers for one item, it will require four times as much work to communicate than if there were one (besides introducing additional variation).

Complexity Introduced by the Degree of Difficulty

A fifth factor introducing complexity is the degree of difficulty of a task. Some of the requirements may be contradictory, such as when a baseball pitcher is told, "Don't walk the batter—but don't give him anything good to hit." Often a task requires a technology still being developed.

DOCUMENTING PROCESS-RELATED EXPECTATIONS

One illustration of the impact of complexity and level of work has been documented by F. Timothy Fuller.* The complexity issue that he dealt with was the complexity introduced in assembling printed circuit boards when not all of the parts are available. The flow diagram of this process is shown in Figure 9.2.

* Fuller, F.T. "Eliminating Complexity From Work: Improving Productivity by Enhancing Quality," *National Productivity Review*, Autumn 1985, pp. 327–344.

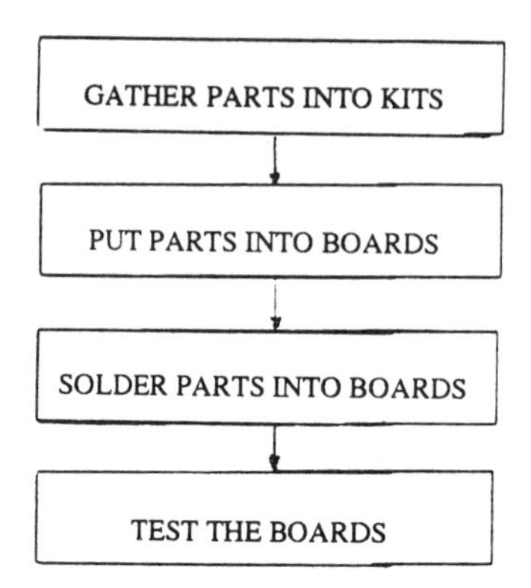

Figure 9.2 Flow diagram of the process of manufacturing circuit boards.

Data showed that when kits were formed, about 98% of the parts were available. However, since there were as many as 100 different components per board, about 75% of the kits had at least one missing part.

Production management considered this to be less than desirable and worked with the Materials group to achieve a lower number of backorders. In the meantime, though, they would perform a little wizardry of their own. When parts were missing, they would load what they had and then store the incomplete boards until the missing parts arrived. When the parts arrived, they would then bring back the partially loaded boards and complete the loading.

The logic was: because the boards were previously processed as far as possible, when the missing parts arrived the boards could quickly be completed.

As luck would have it, orders increased and the demand for additional boards went up. This required borrowing people from another division, and it happened that these were people who were expecting the best. After a couple of weeks, they were making comments like "We don't like it here. Things are too disorganized, and every time we start something, we run out of parts and have to find something else to work on. We didn't have these problems in our other job." This is an example of people as active contributors to the work process.

Fortunately, assembly management was perceptive and decided not to start any kits unless they were complete. The results were remarkable and reinforce the contributions made within the quality triad: the amount of labor to assemble a kit was cut nearly in half, and the time to build a kit decreased from 16.5 days to 5.5 days.

The difference between these two methods was the complexity introduced by partially completed kits. This complexity introduced the following extra duties: identifying which boards were partially completed; inspecting to see which components were missing; finding space for storing partially completed boards; developing systems to keep track of the work in process; keeping a "hot list" of all the parts that were missing; identifying and retrieving incomplete boards once the missing parts arrived; and disrupting people from their normal duties to put in the missing parts.

Doing all these activities showed our perceived wizardry. We did prove our resourcefulness, but we proved it with additional labor to do the job, with additional space to store the incomplete boards, and with the additional procedures, systems, and computers to keep track of the incomplete boards.

SUMMARY

A fourth role closely associated with the Quality Triad is the wizard. The wizard distinguishes himself or herself from the other roles through a low expectation and the mythical perception of a magical ability to overcome any waste successfully.

Waste can be defined as anything that takes resources and contributes nothing. There are three major types of waste: errors, complexity, and uneven levels of work.

Waste can be introduced by suppliers, by the present process, or by customers.

10

Expectations and Responsibilities of the Craftsman

The expectations of a craftsman support the beliefs that there should be a harmony between the craftsman and the process; that there should be a self-involvement with the work; and that there is an ugly and a beautiful way of doing everything. The craftsman expects to have mastery over what is produced and realizes that the source of most errors is the lack of self-control. It is for this reason that the craftsman expects to be in a state of self-control.

EXPECTATION 1: THE CRAFTSMAN SHOULD BE IN A STATE OF SELF-CONTROL

If self-control is not absent, alarm signals should sound, because problems may be occurring. Three conditions were listed earlier as conditions for self-control. They were (a) knowledge of what to do; (b) knowledge of how you are doing; and (c) knowledge of what to do if what you are doing does not agree with what should be happening.

Table 10.1 should help determine how completely the criteria for self-control are met.

RESPONSIBILITY 1: THE CRAFTSMAN SHOULD GIVE MAXIMUM EFFORT FOR SELF-CONTROL

The craftsman realizes that he or she is not a passive observer to achieving self-control. The craftsman realizes that he or she must make the maximum effort to achieve the state of self-control.

Being in a state of self-control is not a natural state. It is an achievement. There will be some situations when we may not meet the three requirements for self-control. It may be that the feedback loop for what we are doing is too long for immediate control, or it could be that the corrective action is unknown. In these situations it should be remembered that the success is just as much in the journey as it is in the destination. If these situations occur, attention should be directed toward making the maximum effort to be in control rather than focusing on the issue of not being in control. Self-control is something we may not be able to achieve all the time. On the other hand, it is possible to focus on our efforts to achive self-control. In this way, one can put the energy into the effort of achieving self-control rather than the mere concern for achieving self-control.

For the craftsman, it is not just a matter of knowing what he or she should be doing. It is a matter of seeing the beauty in what he or she does. It does not matter whether the job is polishing a pair of shoes or taking a spot out of a shirt or designing an operating system for the next generation of computers. The craftsmen that I know realize that there is a beautiful way of doing everything. They realize that there is beauty that goes beyond just knowing how to do something. They realize that it is just as important to recognize what looks good, and why it looks good, as it is to recognize how to do it.

Table 10.1 Levels of Self-Control

Knowledge of What to Do

Level 1. Knowing the criteria for evaluating work
You know the criteria for evaluating your work.
You know the criteria that are most important.
There are no conflicts in criteria.
You know whom to consult in doubtful cases.

Level 2. Knowing how to accomplish the work
You have instructions for what you do.

Level 3. Knowing why something is required
You have seen your work used by others.
You know the consequences of not meeting the criteria.
You review the criteria with your customer from time to time.

Knowledge of How You Are Doing

Length of time for feedback to occur

Level 1. No ability to check your own work, but regular feedback from others

Level 2. Ability to check your own work, but feedback is delayed

Level 3. Feedback as the work is done

Explicitness of feedback

Level 1. Qualitative data on acceptability
You know only if a work meets or fails to meet criteria.

Level 2. Quantitative data on acceptability
You know quantitatively the quality of the work.

Knowledge of How to Make Corrections

Level 1. Knowing when to stop
You know that type of changes you can make.
You know what type of changes need approval before being made.

Level 2. Knowing whom to consult

Level 3. Knowing what changes to make

Those who write job instructions are familiar with the need for explaining the "what" and the "why" of the situation. They realize that by giving this information the task being taught is more easily remembered.

EXPECTATION 2: THE CRAFTSMAN EXPECTS NO WASTE

The craftsman maintains the role by not letting the wizard appear. The craftsman deserves the best and expects the best. This means that all the types of waste (errors, undue complexity, erratic levels of work) raise an alarm whenever encountered.

RESPONSIBILITY 2: THE CRAFTSMAN SHOULD GIVE FEEDBACK TO ELIMINATE WASTE

The craftsman has the matching responsibility to identify the sources of the waste. This is the first step to resolving the problem of waste.

EXPECTATION 3: THERE SHOULD BE SOLUTIONS TO ENDURING WORK PROBLEMS AND ONE SHOULD LEARN AND DO NEW THINGS

I can be very tired after a day of work, but generally it is not the work that tires me. What tires me are all those things that prevent me from accomplishing my work. In fact, the days that I am the most tired are most likely the days I was the least productive. There is a strong linkage between the work "out there" and the person "in here."

The responsibility matching this expectation is the following.

RESPONSIBILITY 3: THERE SHOULD BE CONTINUAL PROCESS IMPROVEMENT

We do not live in a static society. Our society is driven by the desire for more of everything—more knowledge, goods, and power. This results in a society of continual change. The black-and-white television sets that were first introduced have now been replaced by color television sets, which will themselves be replaced eventually by high-density monitors, which too will in time become obsolete. What was considered the best grade of product will be replaced by a new and better product. The same is true for the craftsman. If excellence is to be maintained, a strategy for improvement is required.

There are three kinds of action that can be taken as a result of any failure experience: quick fix, corrective action, and recurrent control action. The quick fix is the shallowest level of fix and the most common kind. Sorting, scrapping, reworking of existing products, temporary modifications, and adjustments to process procedures are examples of the quick fix. These efforts are self-healing in that the need for the activity stimulates effort to fix it. This type of fix represents little quality improvement, and nothing is learned from the failure incident.

The second level of fix, corrective action, is the improvement of the design or production process. This reduces the chance of the same failure occurring with similar items. This type of fix is also shallow, but it is quick and inexpensive. Most manufacturers today have systems in place for handling corrective action.

The deepest level of fix derives the maximum benefit from the failure incident, not only for the current quality deficiency but also for other projects. Here the basic cause of the failure is studied and related to the process that contributed to the cause.

One example in which the cause of failure was pursued to the root is the investigation into the explosion of the space shuttle Challenger. The explosion occurred because of a failure of the field joints of the rocket boosters at low temperature. One first-level fix would be to restrict the ambient temperature at take-off. In this particular flight, that may have prevented the explosion. The explosion may not have occurred also had it not been for the wind shear that twisted the rocket booster.

A deeper fix would be to modify the design of the booster. For example one could

1. Change the O-ring material in the joint,
2. Install joint heaters so that the temperature of 80° could be maintained,
3. Use a secondary seal on the aft field joints,
4. Orient the carbon plies on the exhaust nozzle differently to reduce susceptibility to erosion from exhaust gases,
5. Use custom shims in all field joints to force the joining of the case segments to have the same ovality, resulting in a uniform pressure on the joint O rings.

NASA did not stop here but probed deeper. They explored why this problem was not detected by the testing process. NASA asked what was wrong with the design and testing phases that this problem went undetected, or if it was detected, why this design continued to be used.

Proceeding to the root cause of a problem provides leverage, because the improvement effort will not just affect the units that are to be launched on a single day, nor will it just affect the units produced under a new design. This fix will affect all units designed in the future.

SUMMARY

The Role: A craftsman is one who exercises with skill a mastery of his or her art or profession.

Beliefs: A craftsman believes that there is a harmony between craftsman and process; that there is a beautiful and ugly way of doing everything; that there is a self-involvement (and identification) with the process.

Expectations: A craftsman should exercise self control—knowledge of what should be done; knowledge of what one is doing; knowledge of how to correct discrepancies.

There should be no waste—no errors, no complexity, no unstable workload; there should be solutions to enduring problems.

Responsibilities: A craftsman should give maximum effort for self control; give feedback to the source of waste; make continual process improvements, using process leverage to find root-cause solutions.

11

Expectations and Responsibilities of the Connoisseur

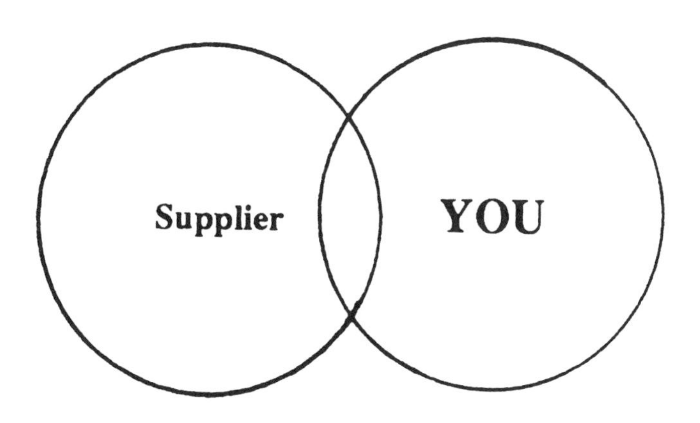

For connoisseurs, expecting the best means receiving what is needed to achieve the goals and fulfill themselves. Connoisseurs have three expectations.

EXPECTATION 1: SUPPLIERS SHOULD NOT INTRODUCE ANY WASTE, COMPLEXITY, OR INSTABILITY INTO THE WORK

Suppliers are process partners. They are extensions of the processes of the craftsman. They are to help the work be done more easily and more productively. Achieving higher productivity can only occur if waste, complexity, and erratic levels of work are not introduced into the process. This means that the product or service as it is received should be able to be used successfully.

EXPECTATION 2: SUPPLIERS SHOULD PROVIDE ASSURANCE THAT THE JOB WAS DONE CORRECTLY

The supplier should not introduce waste, complexity, or instability into the process and give assurance that there will be no problem. One form of assurance may be that of a guarantee or a certificate of conformance. Or it may be in the form of statistical data showing the uniformity or consistency of the process of manufacture.

EXPECTATION 3: SUPPLIERS SHOULD HAVE A PROGRAM FOR CONTINUAL PROCESS IMPROVEMENT

One of the responsibilities of the craftsman is to be working for continual process improvement. The work of the supplier is an extension of the work of the craftsman. The same expectations must apply.

There are matching responsibilities for the discriminating buyer.

RESPONSIBILITY 1: THE REQUIREMENTS OF THE PRODUCT OR SERVICE BEING PURCHASED SHOULD BE CLEARLY DEFINED.

Have you ever walked into a beauty shop or barber shop, sat down in the chair, and said, "I would like a quality haircut"? Chances are you have not. When it comes to having our hair cut the strategy of "When I see it, I'll know it" can be employed. But it has certain risks.

The same is true in the relationship between supplier and buyer. If the supplier is to provide what the connoisseur needs, there has to be a clearly defined set of requirements. Without this, errors, complexity, or delays in service will occur.

RESPONSIBILITY 2: FEEDBACK SHOULD BE GIVEN ON HOW WELL THE SERVICE OR PRODUCT WORKED

Have you ever fixed something that you did not know to be broken? Probably not. Chances are that your suppliers will not fix problems associated with the product or services that they supply if they do not know about them. Without feedback on the usage of the product, the supplier may never know it is wrong. Even if the supplier is evaluating the product, his experience may be significantly different from yours.

RESPONSIBILITY 3: SUGGESTIONS FOR IMPROVEMENT SHOULD BE PROVIDED

Suggestions for improvements from our customer are as critical as feedback on the usage of the product. According to Tom Peters, author of *In Search of Excellence*, there have been over 80 studies done to uncover the source of ideas for improvements.* In every one of these, the source of ideas is the customers. It does not matter what business it is. Whether it is insurance, banking, or aircraft engines, the reality of the situation is that the customers account for over 80% of all improvements.

SUMMARY

The Role: A connoisseur is one who enjoys with appreciation and discrimination the subtleties of the work or services provided.

Beliefs: A connoisseur believes that a working relationship is a harmonious, constructive, mutually beneficial partnership; that there is no such thing as a commodity; that suppliers are an extension of the processes and contribute to one's work; that customers' needs are preeminent.

* Peters, T. *Six Keys to Success* (audio tapes), Tape #7, 1980.

Expectations: A connoisseur expects what is needed to achieve goals: that the supplier should not knowingly introduce any waste into the process (errors, complexity, or erratic level of work); that the supplier should provide peace of mind that the job was done correctly; that the supplier should continually work for process improvement.

Responsibilities: The connoisseur should provide what is needed to do the work, defining the requirements of the product or services; giving feedback on usage; giving suggestions for improvement.

12

Expectations and Responsibilities of the Entrepreneur

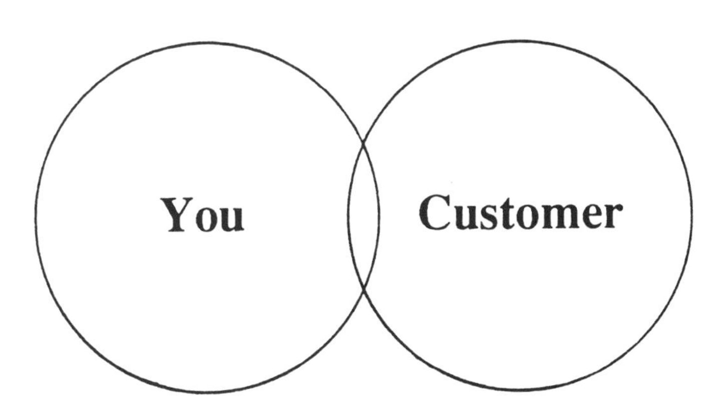

In order for the entrepreneur to fulfill his or her role, three expectations must be met.

EXPECTATION 1: THE REQUIREMENTS OF THE CUSTOMER ARE DEFINED

The entrepreneur expects the best. Expecting the best in this relationship requires that the criteria for self control be met. The first requirement is knowing what the customers' requirements are. Without this knowledge the customers' requirements may not be met. Included in this knowledge is the importance and hierarchy of those requirements to the customer and knowing the impact upon the customer for not meeting requirements.

EXPECTATION 2: THERE IS FEEDBACK EVALUATING THE RESULTS OF THE PRODUCTS OR SERVICES

Although the entrepreneur may think that he or she is having no problems, there is no substitute for feedback. Even though a product or a service is planned well and the plan is well-executed, there are many possibilities for mistakes. Without feedback there is uncertainty and a lack of peace of mind that the expectations are being met. In the extreme case this may be the situation between getting feedback and getting no feedback. In a better situation it is an issue of the delay between producing the product or service and getting feedback on the usage of the product or service.

EXPECTATION 3: THERE IS FEEDBACK FOR IMPROVEMENT

Feedback for improvement is not required just as a technique for knowing what needs to be changed. It is also needed to avoid burnout. Being an entrepeneur takes a lot of energy and resources to be interested in what the customer wants and to creatively satisfy the customers' needs. One way to avoid burnout is feedback from the customer, the person who uses the product or services that you provide. Sometimes the feedback may be an important technical contribution, other times it will be positive psychological support. Associated with these expectations are mutual obligations or responsibilities.

RESPONSIBILITY 1: ONE SHOULD DISCOVER THE NEEDS OF THE CUSTOMER

There is a reality associated with this responsibility. The reality is that the customer measures quality in his or her own terms. Or perhaps a closer reality is that the customer defines quality in his or her own unique, idiosyncratic, human, emotional, end-of-the-day, irrational, erratic terms. If the request of a customer appears to be ridiculous, that may be just the indication of a unique opportunity for discovering something truly innovative.

RESPONSIBILITY 2: ONE SHOULD ASSURE THE CUSTOMER THAT THE JOB IS RIGHT AND ARRANGE FOR FEEDBACK

With all the competing work that must be done, it may be very easy to think that no news is good news. The fact is that one of the surest signs of a bad or declining relationship is the absence of complaints from the customer. Nobody is every completely satisfied, especially over a long period of time. The customer is either not being open or not being contacted. This lack of feedback is both a symptom and a cause of trouble. The entrepreneur arranges for feedback from the customer.

RESPONSIBILITY 3: ONE SHOULD ARRANGE FOR FEEDBACK FOR IMPROVEMENT

The customer is an excellent source of suggestions for improvement, yet sometimes it is hard to get the feedback. If asking for feedback does not generate any ideas, one alternative would be to get reconfirmation or denial of your perception. For example, if you are having difficulty getting feedback, you might say: "As far as I can determine, the work I did for you was perfect in every way. Right?" If this is the case, you may have a very loyal customer.

SUMMARY

The role: An entrepreneur is one who creates customers and builds their loyalty.

Beliefs: An entrepreneur believes that the working relationship is a harmonious, constructive, mutually beneficial partnership; that customers' needs are preeminent; that there is no such thing as a commodity.

Expectations: The entrepreneur expects defined requirements with notice of any changes; feedback on usage; suggestions for improvement.

Responsibilities: The entrepreneur should not knowingly introduce waste into customers' processes; provide assurance that the job is right (peace of mind); make continual process improvement.

Part III

THE I.C.A.R.E. PROCESS

13

Caring About Quality

THREE ROLES, ONE ATTITUDE

The Quality Triad represents three roles, but it does not require a split personality. Each role performs different activities and has different expectations and responsibilities, but there is a common thread that runs through all the roles. This common thread is the attitude that people express in their behavior. It is one of caring.

The entrepreneur, for example, cares enough to listen actively to what the customer wants; he or she cares enough to provide assurance that the product or service is right, and he or she cares enough to offer advice on how the product or service could be used.

The connoisseur cares enough to specify the proper requirements. He or she also cares enough to provide feedback on the usage of the product or service and provide suggestions for improvement. If he or she did not care, he or she would not have even noticed when something was wrong. Rejecting carelessness is a form of caring.

As for the craftsman, addressing the root cause of a problem rather than merely making a quick fix shows caring. Likewise, working toward self-control and listening to the process is only done by a person who cares. A person who did not care would not make the effort to look, to listen, and to work actively to make things happen.

This linkage between caring and quality can be hard to identify, because attitudes are not directly visible and can only be seen through behavior. Yet it is the caring attitudes that produce the right thoughts that produce the right actions that result in quality. Caring is that internal aspect that results in quality that everyone can see.

WHAT IS CARING?

When one is not dominated by feelings of separateness from what one is doing, then one can be said to care about what one is doing. The craftsman, connoisseur, and entrepreneur have this caring. There is a self-involved reality that does not separate them from their work. They take interest and pleasure in the work they do. They participate rather than show passive concern.

THE CARING PARADOX

Paradoxes are statements that seemingly are contradictory or opposed to common sense, yet perhaps are true. Paradoxes surround us. Here are some examples: The less hair you have, the longer it takes to comb. The person one loves most is the person capable of arousing the greatest anger. The harder one works, the more enjoyable is play. There is wisdom in frivolity and frivolity in wisdom.

A caring attitude and the resulting self-realization are not independent. They are related in a paradoxical way summarized in the caring paradox: Self-realization is possible only through caring.

One example of this can be seen in roles within the Quality Triad. Each of the roles shares the caring attitude, but each fulfills and realizes the fulfillment of the role played. The entrepreneur creates a business by creating and keeping a (satisfied) customer, the craftsman excels in his/her craft, the connoisseur is able to excel by standing on the shoulders of giants.

This connection between caring and excellence can also be seen on a different level. In the business world, there is a continual desire to find out what makes companies excel. The book *In Search of Excellence* interprets the results of such a study. It may come as no surprise that Tom Peters, the author, summarizes his findings as: "...this book,...if it is anything, is about caring and commitment."*

CREATING THE CARING PROCESS—I.C.A.R.E.

Caring is made up of caring feelings and the ability to act upon them. Initial sections of the book looked at the beliefs, expectations and attitudes. This section covers a process for care and provides some tools to act with. One tool for putting those caring feelings in action is summarized in the mnemonic acronym, I CARE shown below:

THE "I.C.A.R.E." PROCESS

Identify your Quality Triad

Create the Expectations

Accept your Responsibilities

Reciprocate Cooperation

Enjoy the Benefits

*Peters, T.J. and Waterman, R.H., Jr., *In Search of Excellence*, New York: Warner Books, 1982, p. ix.

14

Identifying the People and Processes in Your Quality Triad

Some of the time, it will be obvious what the process is and who the suppliers and customers are for that process. Other times it may not be as obvious. This section will provide tools and an example for identifying and describing the processes, suppliers, and customers.

Most jobs have associated with them many products or services. In my job as a statistical process control (SPC) manager, for example, there are several products or services that I provide. The most significant are the following:

1. Coordination of the annual SPC planning process
2. Feedback on the execution of plans vs. objectives
3. Training for SPC implementation
4. Communication of the SPC program to customers and suppliers

Associated with each of these products or services is a process to achieve the desired results and a group of customers and suppliers. Sometimes the customers and suppliers are the same for the processes, but in general the customers and suppliers will differ. One worksheet that is useful in constructing the elements in the Quality Triad is shown in Figure 14.1.

The starting point in creating the elements in the Quality Triad is the job. Choose just one service or product associated with that job. For this example I select Training for SPC implementation. The next step is to identify the process used to produce that product. Two ways of describing this process or any other process are a process flow diagram and a cause-and-effect diagram. Together they summarize my understanding of the process. The flow diagram shows the steps used to achieve a particular result. The cause-and-effect diagram shows the process factors that contribute to achieve a particular result.

PROCESS FLOW DIAGRAM

Figure 14.2 shows a process flow diagram. It describes the process for preparing for a training class in statistical process control. It is read by starting at the top. The first step gives the inputs. Inputs are materials, information, or action that initiate the process. In this case, the input is the list of participants for the workshop along with their job and work area, and the training materials. The symbol used for inputs is an oblong.

An arrow indicates the progression to the next step, which in this case is a decision point. Decision points are represented by diamond-shaped boxes. Decisions are places in the process where a true/false, yes/no, or pass/fail determination must be made.

Decision statements allow looping to occur. Loops are paths that can take you either back to previous process steps or ahead to later process steps. In this case the decision is to select participants based on either the work area or the job description. If the participants are chosen by area, the participants of a work area are selected. Otherwise, the type of job, such as that of an engineer, operator, or supervisor, is chosen.

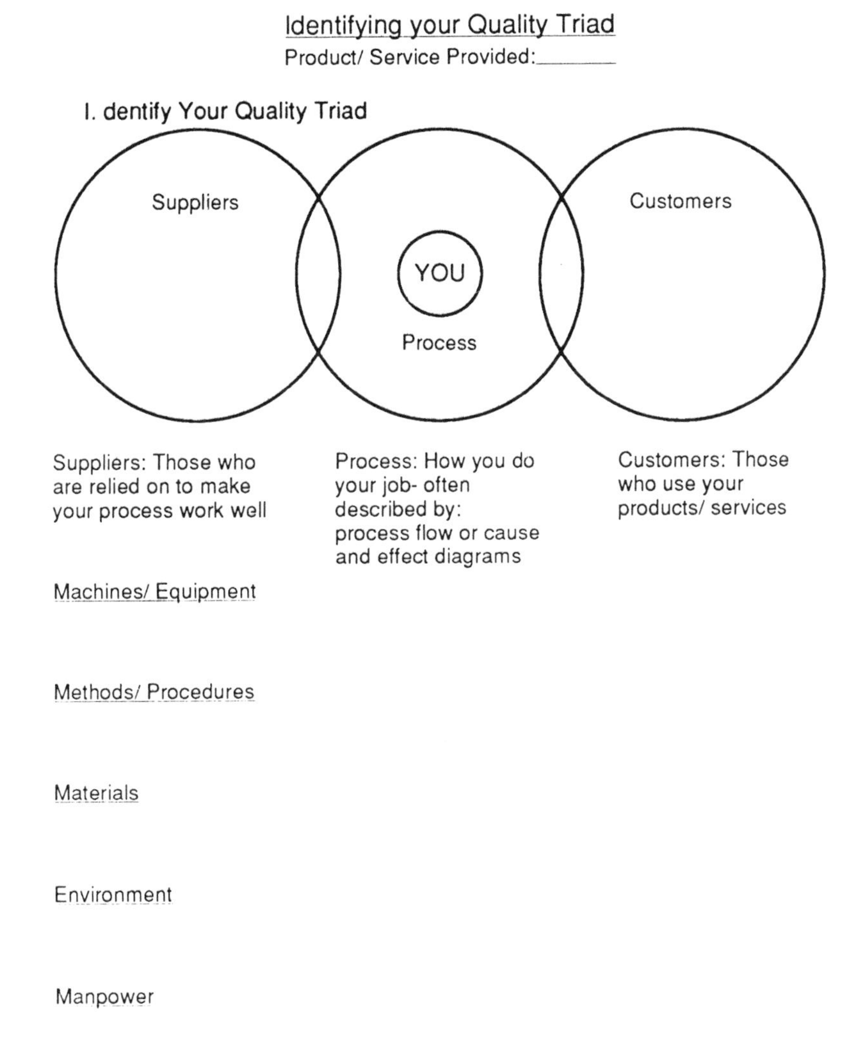

Figure 14.1 I.C.A.R.E. worksheet.

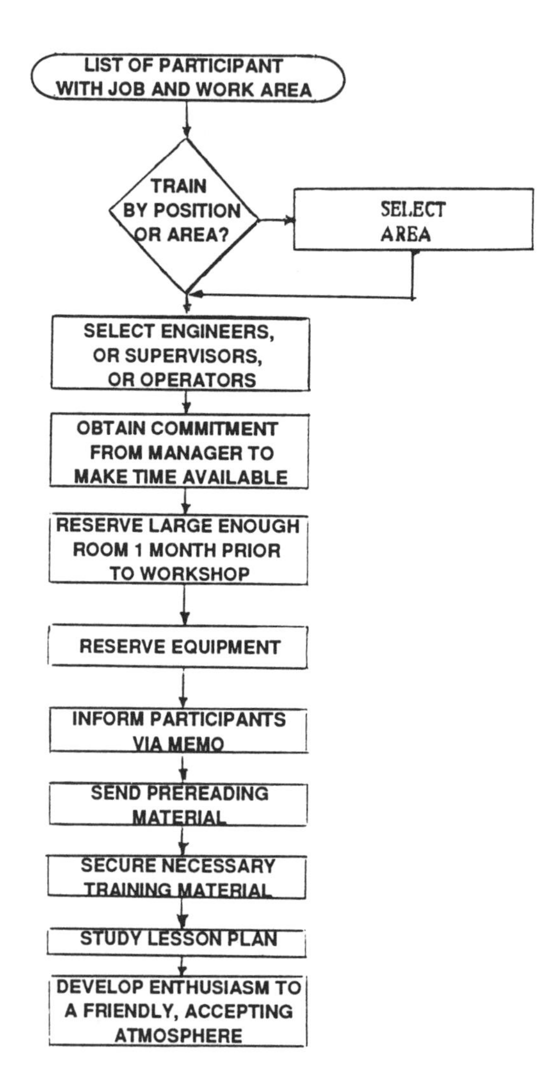

Figure 14.2 A process flow diagram: preparation for a workshop.

In either case, the next arrow flows to a rectangular box. This is an operation step. In this case, the step is to obtain the commitment from the appropriate managers to make time available for the training and follow-up. Further steps are to reserve a room, reserve equipment, invite participants, send prereading material, secure the necessary training material, study the lesson plan, and develop enthusiasm for the class.

Process flow diagrams are used (1) when there is a need to explain a process for information or training purposes, or to document a process; and (2) when an intervention in a process is needed for problem resolution or process improvement.

There are several reasons for using a flow diagram. Such a diagram (1) gives a visual display of the total current process; (2) shows the roles and relationships between steps and people involved in the process; (3) helps explain a process to others for training or informational reasons; (4) indicates problem areas, areas of unnecessary loops and complexity, and areas where simplification of a process is possible; (5) helps to identify where to collect data and investigate further; (6) helps identify what elements may impact our process performance; and (7) helps in documenting and standardizing the process.

CAUSE-AND-EFFECT DIAGRAM

The second tool for describing the process is the cause-and-effect diagram. A cause-and-effect diagram for what causes a good SPC workshop is shown in Figure 14.3. The purpose of this diagram is to show the factors that can cause a particular result. It was developed in Japan by Dr. Kauro Ishikawa in 1952.*

It is closely related to another type of chart that is prevalent in American industry, the organizational charts. Organizational charts create a visual display of reporting relationships within a company. An example is shown in Figure 14.4. It shows that Scott reports to Ed, who reports to Dorothy, who reports to George. The key word that connects

* Ishikawa, K. *What is Total Quality Control the Japanese Way?* Englewood Cliffs, N.J.: Prentice-Hall, 1985, p. 64.

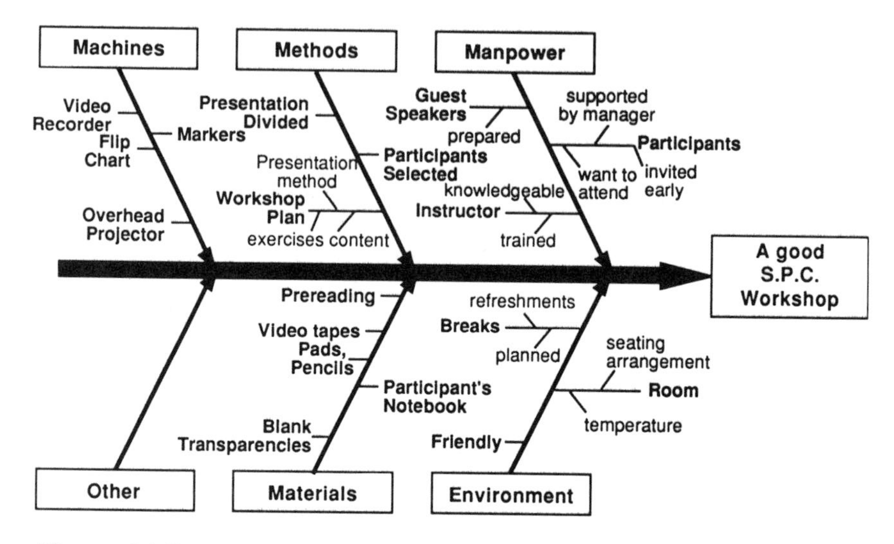

Figure 14.3 A cause-and-effect diagram: preparation for a workshop.

the people is the word, or operator, "reports to" (not to be confused with "follows the instructions of," or "pays any attention to").

The cause-and-effect diagram creates a similar type of hierarchy by using a different operator. The operator used here is "causes."

One branch of the cause-and-effect diagram in Figure 14.3 would be read, "Management's desire to implement SPC causes managers to support SPC, which causes participants to want to attend, which causes a good SPC workshop."

Dr. Ishikawa lists six main advantages of his diagram.

1. Making a cause-and-effect diagram is educational in itself.
2. A cause-and-effect diagram is a guide for discussion.
3. The causes are sought actively and the results are written in on the diagram.
4. Data are collected with a cause-and-effect diagram.

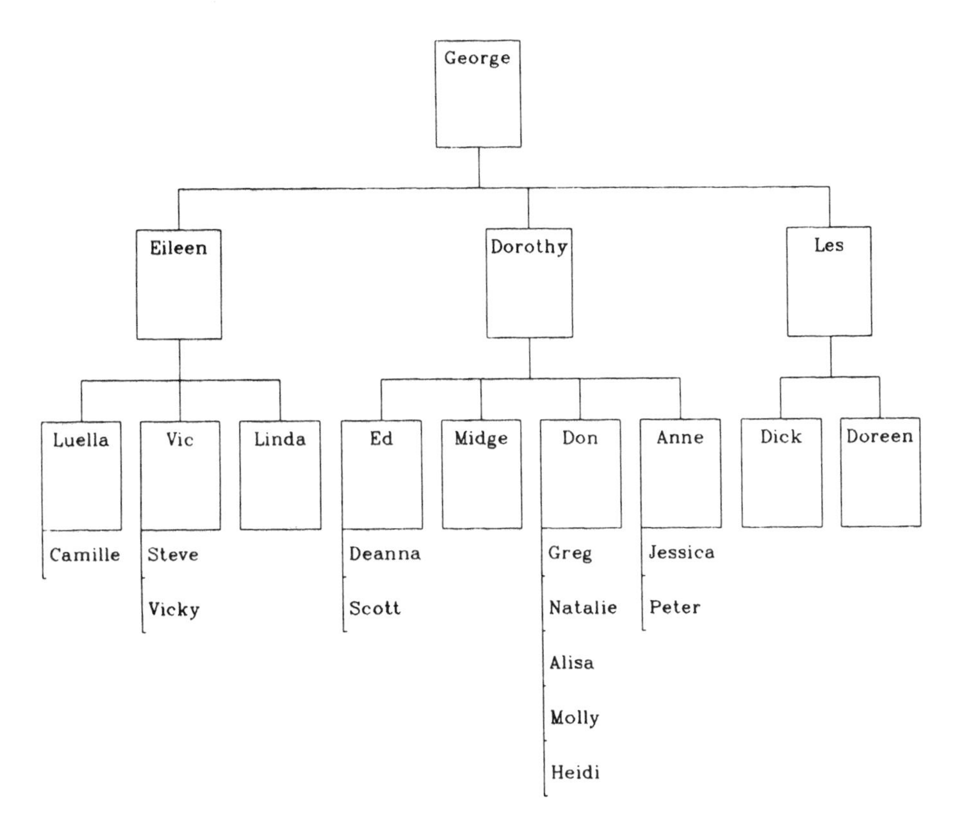

Figure 14.4 An organizational chart: reporting relationships.

5. A cause-and-effect diagram shows the level of technology.
6. A cause-and-effect diagram can be used for any problem.

IDENTIFYING THE CUSTOMERS AND SUPPLIERS

Once a process flow diagram and a cause-and-effect diagram have been created to summarize your understanding (craftsmanship) of the pro-

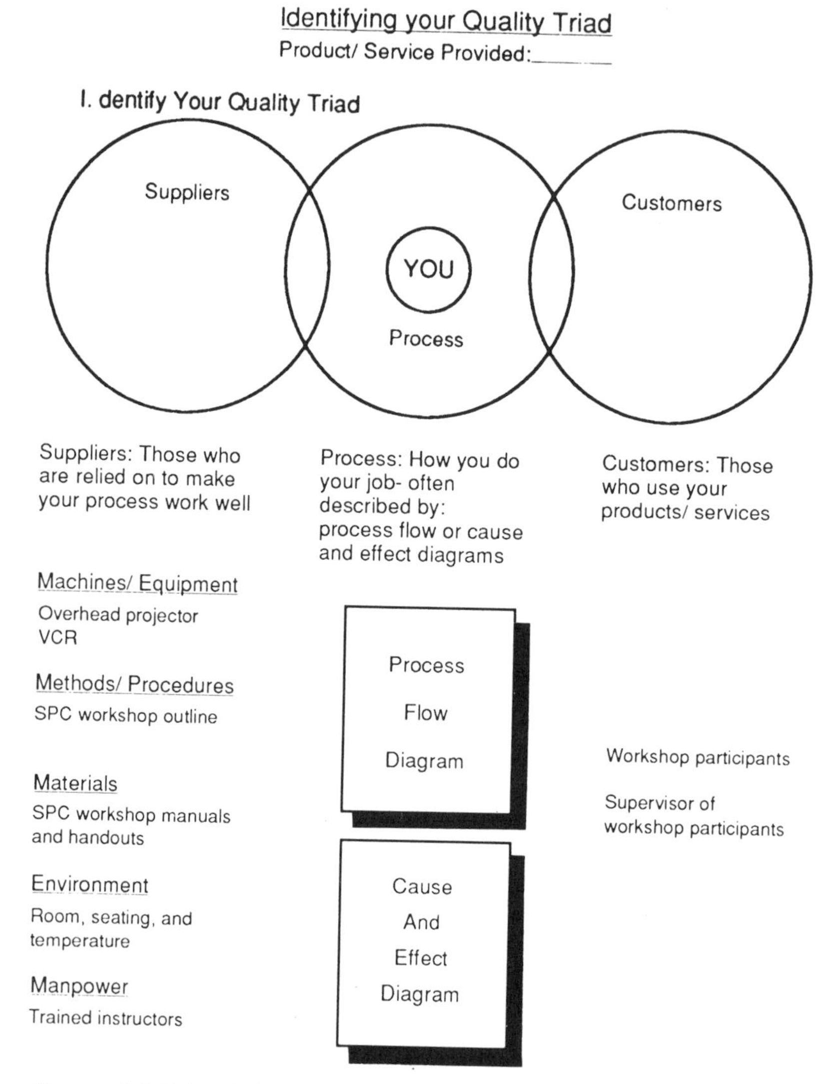

Figure 14.5 Partially completed I.C.A.R.E. worksheet. Shadows behind the boxes indicate that there is another level of detail. In this case, there is a separate process flow diagram (page 128) and a cause-and-effect diagram (page 130).

cess, the suppliers and customers can be identified. The customers are those who use the outputs of the process. In this case they are workshop participants and their managers. The suppliers are those who are relied on to make the process work well. In this case, the following supplies are needed (and associated with each supply is a supplier): equipment (overhead projector, VCR), methods (SPC workshop leaders course outline), reading, materials (workshop manuals and handouts), environment (room, seating, temperature), and manpower (trained instructors).

The worksheet to this point is shown in Figure 14.5.

SUMMARY

The process can be identified by first identifying the product and service provided. It can be described by two diagrams, the process flow diagram and the cause-and-effect diagram. The process flow diagram shows the steps and the time relationship to achieve a result. The cause-and-effect diagram shows the factors that contribute to achieve that same result.

The customers are the people who use or rely on the products or services provided. The suppliers are identified by first identifying the resources or supplies that are needed to let the process run efficiently and then finding the people who can provide them.

15

Creating the Expectations

The expectations associated with the craftsman, connoisseur and entrepreneur were established as a minimum set in order for the person to accomplish what was expected in that role. Worksheet 15.1 shows the expectations associated with three roles.

For example, the expectations of the craftsman are: self control, no waste and solution to enduring problems. These are listed on worksheet 15.1 under the craftsman role. Worksheet 15.2 takes the general expectations of the craftsman and translates the general expectations into specific, operational expectations associated with a specific process. In this case it would identify these expectations associated with a SPC workshop. They are listed under "Fulfillment of Expectations."

Worksheet 15.3 is associated with the connoisseur. Worksheet 15.4 is associated with the entrepreneur. A sample of the completed worksheets for Creating the Expectations is shown in worksheets 15.5–15.12.

Worksheet 15.1 Expectations associated with three roles. The drop shadow indicates that there is additional detail that cannot be shown on this worksheet. Supporting detail is given in Worksheets 15.2–15.4.

Creating the Expectations

Product/ Service Provided:______

I. dentify Your Quality Triad

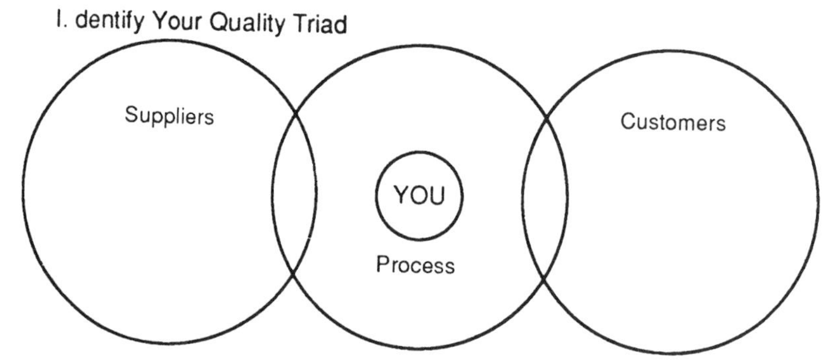

C. reate the Expectations

Assurance it is right	Self-control	Defined requirements
Guarantee/ warranty Product test data Process data	Knowledge of what you are supposed to do Feedback on how you are doing Ability to change	Know requirements Know product usage Know consequences of nonconformance
No supplier induced waste	**No process induced waste**	**Feedback on product**
No errors No complexity Level workloads	No errors No complexity Level workloads	Timely feedback Descriptive feedback
Continual improvement	**No enduring problems**	**Suggestions for improvement**

Worksheet 15.2 Specific expectations of the craftsmen.

Creating the Expectations

for the Craftsman

Product/Service provided:________________________________

Craftsman's Expectations	Fulfillment of Expectations
Self-Control	
Knowledge of what you are supposed to do	
Feedback on how you are doing	
Ability to regulate	
No process-induced waste	
No errors	
No complexity	
Level workload	
No enduring problems	

Worksheet 15.3 Specific expectations of the connoisseur.

Creating the Expectations

for the Connoisseur

Product/Service provided: ______________________________

Supplier: ______________________________

Connoisseur's Expectations	Fulfillment of Expectations
Assurance it is right	
Guarantee or warranty	
Product test data	
Process data	
No supplier-induced waste	
No errors	
No complexity	
Level workload	
Continual improvement	

Worksheet 15.4 Specific expectations of the entrepreneur.

Creating the Expectations

for the Entrepreneur

Product/Service provided: ______________________________

Customer: ______________________________

Entrepreneur's Expectations	Fulfillment of Expectations
Defined requirements	
Know requirements	
Know product usage	
Know consequences of nonconformance	
Feedback on product	
Timely feedback	
Descriptive feedback	
Suggestions for improvement	

Worksheets 15.5–15.12 Samples of a completed worksheet by people involved in a successful SPC workshop.

Creating the Expectations

for the Craftsman

Product/Service provided: SPC workshop

Process: Preparation for SPC workshop

Craftsman's Expectations	Fulfillment of Expectations
Self-Control	
Knowledge of what you are supposed to do	SPC leader's guide for SPC workshop
Feedback on how you are doing	Workshop evaluations
Ability to regulate	
No process-induced waste	
No errors	No mistakes in handouts
No complexity	
Level workload	Workshop does not conflict with other obligations
No enduring problems	No known problems in workshop not being worked on

Worksheet 15.6

Creating the Expectations

for the Connoisseur

Product/Service provided: SPC workshop

Supplier: Overhead projector, VCR

Connoisseur's Expectations	Fulfillment of Expectations
Assurance it is right	
Guarantee or warranty	
Product test data	Equipment tested to verify it works prior to delivering
Process data	
No supplier-induced waste	
No errors	Equipment set up properly
No complexity	Backup equipment made available
Level workload	
Continual improvement	New equipment evaluated periodically

Worksheet 15.7

Creating the Expectations

for the Connoisseur

Product/Service provided: SPC workshop

Supplier: SPC leader's guide

Connoisseur's Expectations	Fulfillment of Expectations
Assurance it is right	
Guarantee or warranty	
Product test data	Workshops successfully given using the leader's guide
Process data	
No supplier-induced waste	
No errors	Known errors corrected
No complexity	Known complexity reduced
Level workload	
Continual improvement	Leader's guide updated regularly

Worksheet 15.8

Creating the Expectations

for the Connoisseur

Product/Service provided: SPC workshop

Supplier: Workshop manuals and handouts

Connoisseur's Expectations	Fulfillment of Expectations
Assurance it is right	
Guarantee or warranty	
Product test data	
Process data	Clearly defined process for producing workshop manuals and handouts
No supplier-induced waste	
No errors	
No complexity	Process has been studied for complexity
Level workload	
Continual improvement	Handouts updated as new examples are developed

Worksheet 15.9

Creating the Expectations

for the Connoisseur

Product/Service provided: SPC workshop

Supplier: Room seating and temperature

Connoisseur's Expectations	Fulfillment of Expectations
Assurance it is right	
Guarantee or warranty	
Product test data	
Process data	Standardized process to make sure the room is set up correctly
No supplier-induced waste	
No errors	No double booking of room
No complexity	Signing up for a room is easily done
Level workload	
Continual improvement	Suggestions for improvements evaluated and incorporated

Worksheet 15.10

Creating the Expectations

for the Connoisseur

Product/Service provided: SPC workshop

Supplier: trained instructors

Connoisseur's Expectations	Fulfillment of Expectations
Assurance it is right	
Guarantee or warranty	Knowledge of SPC
Product test data	Experience in facilitating workshop
Process data	
No supplier-induced waste	
No errors	No errors in presentation of material
No complexity	
Level workload	
Continual improvement	Suggestions for improvement are evaluated and incorporated

Worksheet 15.11

Creating the Expectations

for the Entrepreneur

Product/Service provided: SPC workshop

Customer: workshop participant

Entrepreneur's Expectations	Fulfillment of Expectations
Defined requirements	
Know requirements	Workshop to provide knowledge of SPC principles, direct application to job, and skill in using SPC tools
Know product usage	
Know consequences of nonconformance	
Feedback on product	
Timely feedback	Participant to perform exercises during workshop
Descriptive feedback	
Suggestions for improvement	Participant to critique workshop

Worksheet 15.12

Creating the Expectations

for the Entrepreneur

Product/Service provided: SPC workshop

Customer: Supervisor of participant

Entrepreneur's Expectations	Fulfillment of Expectations
Defined requirements	
Know requirements	Supervisor tells expectations
Know product usage	Tour of area by supervisor
Know consequences of nonconformance	
Feedback on product	
Timely feedback	Feedback given within 6 weeks of workshop
Descriptive feedback	
Suggestions for improvement	Suggestions given on postimplementation

16

Accepting Your Responsibilities

Associated with each expectation is an obligation. For each expectation, the corresponding responsibility is given here for the craftsman, the connoisseur, and the entrepreneur.

CRAFTSMAN

Expectation	Corresponding responsibility
Operation in self-control	Work toward self-control
No process-induced waste	Reduce waste
Solutions to enduring problems	Continual process improvement

CONNOISSEUR

Expectation	Corresponding responsibility
No supplier-induced waste	Clearly defined requirements
Assurance it is right	Feedback on usage
Continual process improvement	Suggestions for improvement

ENTREPRENEUR

Expectation	Corresponding responsibility
Defined requirements	Do not induce waste
Feedback on usage	Assurance it is right
Feedback for improvement	Continual process improvement

The final process and partnership worksheet is shown in Worksheet 16.1. Following this are the worksheets for the expectations and responsibilities for each of the relationships of the craftsman, the connoisseur and the entrepreneur (Worksheets 16.2–16.4). As in the previous chapter, Worksheets 16.5–16.11 are samples of completed worksheets from a SPC workshop.

Worksheet 16.1 Responsibilities associated with three roles. The next level of detail indicated by the drop shadow is shown on Worksheets 16.2–16.4.

Accepting the Responsibilities

Product/ Service Provided: SPC Training

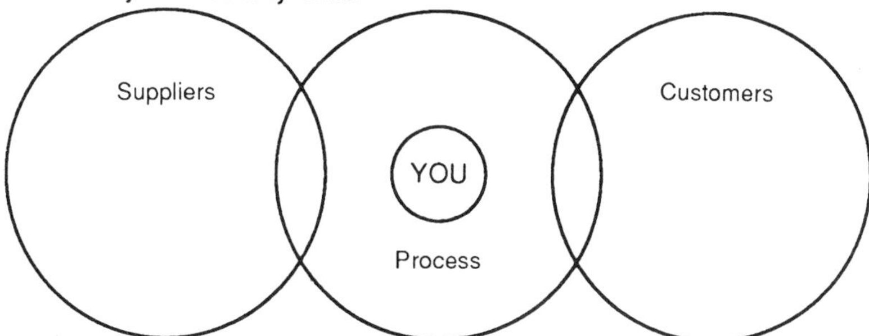

C. reate the Expectations

Assurance it is right	Self-control	Defined requirements
No supplier-induced waste	No process-induced waste	Feedback on product
Continual process improvement	No enduring problems	Suggestions for improvement

A. ccept the Responsibilities

Define wants	Work toward self-control	Assure it is right
Give specifications Explain impact of not meeting specifications	Construct flow diagram Construct cause and effect diagram	Product test data Guarantee
Feedback on product	Reduce waste	Induce no waste
Give feedback	Analyze flow diagram Chart output with time	No errors No complexity Level workload
Suggest improvements	Continual process improvement	Continual process improvement

---

Worksheet 16.2 The craftman's responsibilities.

Accepting the Responsibilities

for the Craftsman

Product/Service provided: ______________________________

Craftsman's Expectations	Fulfillment of Expectations
Self-control	
No process-induced waste	
No enduring problems	

Craftsman's Responsibilities	Fulfillment of Responsibilities
Work toward self-control	
Construct flow diagram	
Construct cause-and-effect diagram	
Reduce Waste	
Analyze flow diagram	
Chart output with time	
Continual process improvement	

Worksheet 16.3 The connoisseur's responsibilities.

Accepting the Responsibilities

for the Connoisseur

Product/Service provided: ______________________________

Supplier: ______________________________

Connoisseur's Expectations	Fulfillment of Expectations
Assurance it is right	
No supplier-induced waste	
Continual improvement	

Connoisseur's Responsibilities	Fulfillment of Responsibilities
Define wants Give specifications	
Explain impact of not meeting specifications	
Feedback on product Give feedback	
Suggest improvements	

Worksheet 16.4 The entrepreneur's responsibilities.

Accepting the Responsibilities

for the Entrepreneur

Product/Service provided: ______________________

Customer: ______________________

Entrepreneur's Expectations	Fulfillment of Expectations
Defined requirements	
Feedback on product	
Suggestions for improvement	

Entrepreneur's Responsibilities	Fulfillment of Responsibilities
Assurance it is right	
Product test data	
Guarantees	
Induce no waste	
No errors	
No complexity	
Level workload	
Continual process improvement	

Worksheets 16.5–16.12 Samples of completed worksheets by people involved in a successful SPC workshop.

Accepting the Responsibilities

for the Craftsman

Product/Service provided: SPC Workshop

Process: Preparation for SPC Workshop

Craftsman's Expectations	Fulfillment of Expectations
Self-control	SPC leader's guide for SPC workshop Workshop evaluations
No process-induced waste	No mistakes in handouts Workshop does not conflict with other obligations
No enduring problems	No known problems in workshop not being worked on

Craftsman's Responsibilities	Fulfillment of Responsibilities
Work toward self-control	
Construct flow diagram	
Construct cause-and-effect diagram	
Plot key indicators	Plot rating from workshop evaluations with time

Continued

Worksheet 16.5 (*Continued*)

Reduce waste	
Analyze flow diagram	Identify rework loops
Chart waste with time	Chart errors with time
Continual process improvement	Observe and learn from other craftsmen

Worksheet 16.6

Accepting the Responsibilities

for the Connoisseur

Product/Service provided: SPC workshop

Supplier: Overhead projector, VCR

Connoisseur's Expectations	Fulfillment of Expectations
Assurance it is right	Equipment tested to verify it works prior to delivering
No supplier-induced waste	Equipment set up properly Backup equipment made available
Continual improvement	New equipment evaluated periodically

Connoisseur's Responsibilities	Fulfillment of Responsibilities
Define wants	
Give specifications	Reserve equipment when needed State any special requirement. Explain unusual requirements
Explain impact of not meeting specifications	
Feedback on product	
Give feedback	Tell how the equipment worked
Suggest improvements	Make any suggestions for improvement

Worksheet 16.7

Accepting the Responsibilities

for the Connoisseur

Product/Service provided: SPC workshop

Supplier: SPC leader's guide

Connoisseur's Expectations	Fulfillment of Expectations
Assurance it is right	Workshops successfully given using the leader's guide
No supplier-induced waste	Known errors corrected Known complexity reduced
Continual improvement	Leader's guide updated regularly

Connoisseur's Responsibilities	Fulfillment of Responsibilities
Define wants	
Give specifications	All the necessary documentation to conduct a good SPC workshop
Explain impact of not meeting specifications	
Feedback on product	
Give feedback	Document sections that have errors or complexity
Suggest improvements	Give suggestions that would improve leader's guide

Worksheet 16.8

Accepting the Responsibilities
for the Connoisseur

Product/Service provided: SPC workshop

Supplier: Workshop manuals and handouts

Connoisseur's Expectations	Fulfillment of Expectations
Assurance it is right	Clearly defined process for producing workshop manuals and handouts
No supplier-induced waste	Process has been studied for complexity
Continual improvement	Handouts updated as new examples are developed

Connoisseur's Responsibilities	Fulfillment of Responsibilities
Define wants	
Give specifications	All materials clearly reproduced, delivered one day prior to workshop
Explain impact of not meeting specifications	Explain space allocation for storing materials

(*Continued*)

Worksheet 16.8 (*Continued*)

Feedback on product	
Give feedback	Give feedback on how well manuals and handouts met expectations
Suggest improvements	Give suggestions that can improve workshop manuals and handouts

Worksheet 16.9

Accepting the Responsibilities

for the Connoisseur

Product/Service provided: SPC workshop

Supplier: Room seating and temperature

Connoisseur's Expectations	Fulfillment of Expectations
Assurance it is right	Standardized process to make sure the room is set up correctly
No supplier-induced waste	No double booking of room Signing up for room is easily done
Continual improvement	Suggestions for improvements evaluated and incorporated

Connoisseur's Responsibilities	Fulfillment of Responsibilities
Define wants	
Give specifications	Reservations show how room is to be set up
Explain impact of not meeting specifications	
Feedback on product	
Give feedback	Tell how the room and environment worked for the workshop
Suggest improvements	Give suggestions for improvement

Worksheet 16.10

Accepting the Responsibilities
for the Connoisseur

Product/Service provided: SPC workshop

Supplier: trained instructors

Connoisseur's Expectations	Fulfillment of Expectations
Assurance it is right	Knowledge of SPC Experience in facilitating workshop
No supplier-induced waste	No errors in presentation
Continual improvement	Suggestions for improvement evaluated and incorporated

Connoisseur's Responsibilities	Fulfillment of Responsibilities
Define wants	
Give specifications	Discuss leader's guide to gain consensus on materials to be covered
Explain impact of not meeting specifications	
Feedback on product	
Give feedback	Give feedback on workshop evaluations
Suggest improvements	Give all suggestions for improvement

Worksheet 16.11

Accepting the Responsibilities

for the Entrepreneur

Product/Service provided: SPC workshop

Customer: Workshop participant

Entrepreneur's Expectations	Fulfillment of Expectations
Defined requirements	Workshop to provide knowledge of SPC principles, direct application to job, and skill in using SPC tools
Feedback on product	Participant to perform exercises during workshop
Suggestions for improvement	Participant to critique workshop

Entrepreneur's Responsibilities	Fulfillment of Responsibilities
Assure it is right	
Product test data	Assurance from past participants
Guarantees	Offer to give additional help, if participant is having trouble
Induce no waste	
No errors	All known errors have been corrected

(*Continued*)

Worksheet 16.11 (*Continued*)

No complexity	Do not introduce superfluous training material
Level workload	Time allowed for participation in workshop
Continual process improvement	Review suggestions for improvement

Worksheet 16.12

Accepting the Responsibilities
for the Entrepreneur

Product/Service provided: SPC workshop

Customer: Supervisor of participant

Entrepreneur's Expectations	Fulfillment of Expectations
Defined requirements	Supervisor tells expectations Tour of area by supervisor
Feedback on product	Feedback given within 6 weeks of workshop
Suggestions for improvement	Suggestions given on post-implementation

Entrepreneur's Responsibilities	Fulfillment of Responsibilities
Assure it is right	
Product test data	
Guarantees	Preworkshop discussion of objectives
Induce no waste	
No errors	Known errors have been corrected
No complexity	

(*Continued*)

Worksheet 16.12 (*Continued*)

Level workload	Workshop scheduled in advance
Continual process improvement	Listen to suggestions for improvement

17

Reciprocating Cooperation in the Partnership

ASSURING A LASTING PARTNERSHIP

The goal of the relationship between the buyer and the seller is mutual benefit. But there is sometimes the temptation to try to achieve benefits from the other partner without investing anything in the relationship. For example, a buyer could provide poorly defined requirements and never work to help clarify them. Then if an issue of acceptability came up, the buyer might try to hold the entrepreneur responsible for correcting the situation. On the other hand, a seller who was not willing to invest time in the relationship might provide limited assurance that the product will work properly and yet rely on the connoisseur for feedback on how the product performed.

Consider the following situation. You are an apple merchant; you buy premium apples and resell them. You have arranged with a supplier to buy his apples for $2.00. You have several customers who will buy them for $3.00, but you have only the one supplier. If you do not buy his apples, the supplier will have to sell them for $1.00, so it is advantageous for both of you to conduct this transaction.

You agree to buy apples daily, but because of conflicts in schedules, you cannot arrange a mutual meeting time. You agree to leave the money in a bag at a designated spot in the woods and to pick up the apples at another designated location. You shake hands on the arrangement and the deal is set.

As you are walking away you realize that there is an inherent risk in this arrangement. The supplier could decide to pick up the bag of money and leave an empty bag. Similarly, you realize you could enhance your profits by leaving an empty money bag. There are four possibilities.

1. You both cooperate. In this case both you and the seller make $1.00. (The profit to you is the difference between the $3.00 for the resold apples and the $2.00 that you paid for them; the seller's profit is the difference between the $2.00 for the apples and the $1.00 that he otherwise would have made.)

2. You cooperate, but your supplier does not. In this situation, you lose the $2.00 you left and receive nothing, so you are out $2.00. The supplier, on the other hand, makes $2.00 and keeps his apples.

3. You do not cooperate, but your supplier does. In this case, you leave no money, yet you collect the bag of apples, which you sell for $3.00. The supplier gets no money and loses the apples that he otherwise could have sold for $1.00.

4. Neither you nor your supplier cooperate. In this case there is no net change for either of you.

These four outcomes as seen by you and the supplier are shown in Figure 17.1.

The outcome for the supplier is better when not cooperating than when cooperating. Also, as shown in Figure 17.2, the outcome for you is better when not cooperating than when cooperating.

SUPPLIER

YOU	cooperates	doesn't cooperate
cooperate	Supplier makes $1.00	Supplier makes $2.00
don't cooperate	Supplier loses $1.00	Supplier breaks even ($0.00)

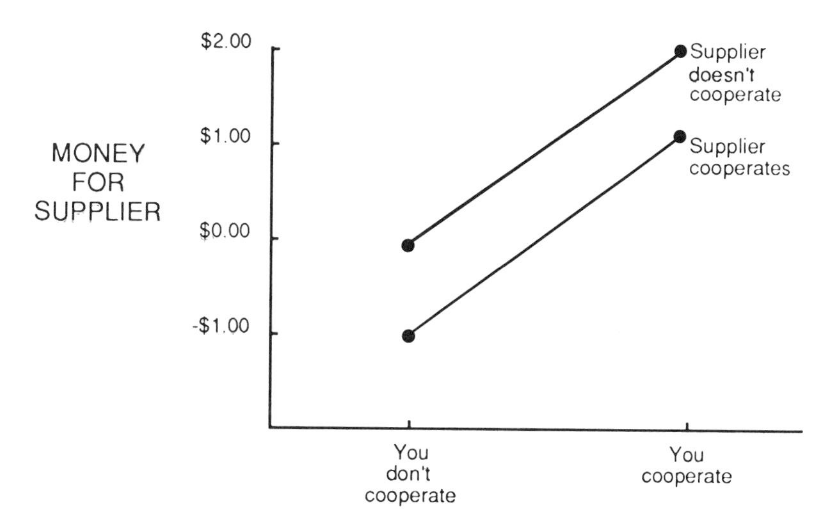

Figure 17.1 The outcome for the supplier is better when not cooperating than when cooperating.

		SUPPLIER	
		cooperates	doesn't cooperate
YOU	cooperate	You make $1.00	You lose $2.00
	don't cooperate	You make $3.00	You break even ($0.00)

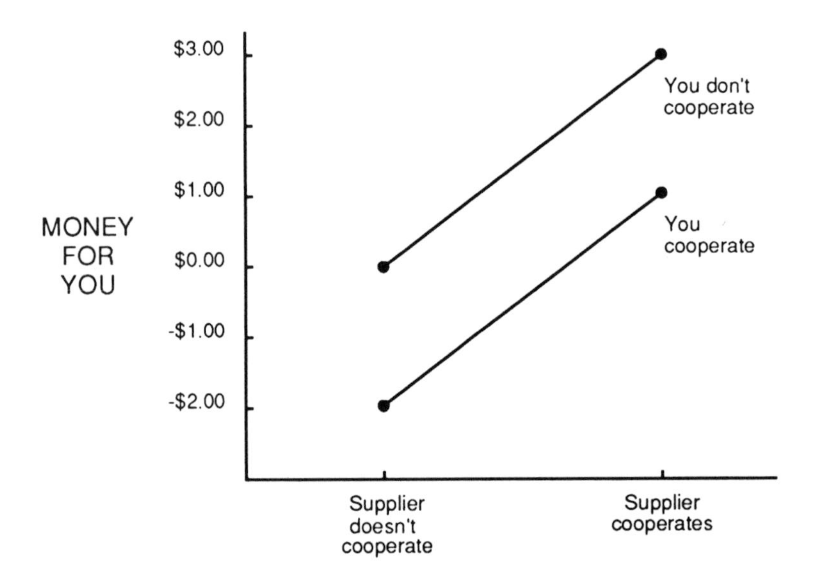

Figure 17.2 The outcome for you is better when not cooperating than when cooperating.

If you examine the tables or the graphs, it is a bit disturbing to see that the supplier stands to make more money by not cooperating, independently of what you do. For the cases where you cooperate, the supplier makes $2.00 if he does not cooperate and only $1.00 if he does. For the cases where you do not cooperate, the supplier loses no money if he does not cooperate, but he stands to lose $1.00 if he cooperates.

The supplier who is equally aware of the situation can see that you likewise stand to benefit more on a single transaction by leaving the bags empty than by cooperating.

If you and your supplier follow this sort of logic, both of you will leave an empty bag, and neither of you will make the mutually beneficial transaction you both agreed to. What then, is the best strategy for optimizing your gains?

This question of when a person should cooperate, and when a person should be selfish in an ongoing interaction with another person, was the basis for a study done by Robert Axelrod of the Department of Political Science and Institute for Public Policy Studies at the University of Michigan at Ann Arbor. He invited experts in game theory to submit computer programs to optimize gains for this situation. The computer programs could use the history of previous transactions to determine whether or not to cooperate. Entries came from game theorists in economics, psychology, sociology, political science, and mathematics. He ran the entries and a random rule (similar to flipping a coin) against each other in a round-robin tournament. Every program played every other program (and a clone of itself) 200 times. The tournament was run five times in a row, so that statistical fluctuations would be smoothed out by averaging.

The program that won was submitted by Anatol Rapoport of the University of Toronto. It was the shortest of all programs submitted. It was five lines long. The strategy was very simple and has been called Tit for tat. It is one in which the first move is to cooperate. Thereafter, the program does whatever the other program did on the previous move. If the program cooperated, the move was to cooperate. If the program did not cooperate, the move was not to cooperate.

PROPERTIES CONTRIBUTING TO SUCCESS

Be Nice and Expect the Best

When Axelrod analyzed what accounted for the relative success of the programs, he discovered that the single property that distinguished the relatively high-scoring entries from the relatively low-scoring ones was the property he calls being *nice*. He used this term for not being the first not to cooperate. Each of the eight top-ranking entries was nice; none of the other entries was. There was a substantial gap in the score between the nice entries and the others. The nice entries received tournament averages of between 472 and 504, while the best of the entries that were not nice received only 401 points. Thus not being the first to defect separated the more successful from the less successful.

Be Forgiving

A second key in the success of a program was forgiveness. Forgiveness can be informally described as the propensity to cooperate after the other program has defected. The winner, Tit for tat, is unforgiving for one move, but thereafter is totally forgiving of that defection. After one punishment, it lets bygones be bygones. Other programs that employed permanent or extended retaliation did not do as well.

Axelrod published the summaries of these findings and decided to hold a larger computer tournament. For this tournament he not only invited all the participants in the first round but also advertised in computer hobbyist magazines to attract people who were addicted to programming and would be willing to devote a good deal of time to working out and perfecting strategy. He described the strategic concepts of niceness and forgiveness that were the lessons of the tournament and also the strategic pitfalls to avoid.

Be Provocable

There was a large response to Axelrod's call for entries. Entries were received from six countries, from people of all ages, and from eight different academic disciplines. The outcome was nothing short of stunning: Tit for tat, the simplest program, won again. Once again the nice programs finished well. Of the top 15, all but one was nice (the one that was not nice finished eighth). Of the bottom 15, only one was nice. The overall lesson of the first tournament was in essence "Be nice and be forgiving." Apparently, however, many people just could not get themselves to believe it. It took the second tournament to prove them wrong. From the second tournament a third key strategic concept emerged, that of provocability—the idea that one should get mad quickly at defectors and retaliate. Thus a more general lesson is "Be nice, be provocable, and be forgiving."

One paradox in the success story of Tit for tat is that it did not defeat a single one of its rivals in their encounters. This is not a quirk; it is in the nature of Tit for tat. It cannot defeat anyone; the best it can achieve is a tie, and often it loses (although not by much). Axelrod makes this point very clear: "Tit for tat won the tournament, not by beating the other player, but by eliciting behavior from the other player that allowed both to do well. Tit for tat was so consistent at eliciting mutually rewarding behavior that it attained a higher overall score than any other strategy in the tournament.

"So in the non-zero-sum world you do not have to do better than the other player to do well yourself. This is especially true when you are interacting with many different players. Letting others do well is fine, as long as you do well yourself. There is no point in being envious of the success of the others, since in a long-term relationship, the success of others is necessary for you to do well."*

Axelrod translated these findings of the tournament into four suggestions for how to do well. They are as follows:

Do Not Be Envious

Work for relationships in which both parties win.

*Hofstader, D.R., "Metamagical Themas," *Scientific American,* 248:5 (May, 1983).

Do Not Be the First to Defect

Both tournaments show that it pays to cooperate as long as the other player is cooperating.

Reciprocate for Both Cooperation and Defection

Tit for tat was able to compete with a wide variety of strategies because it practiced reciprocity. In responding to a defection, Tit for tat responded with like for like, one time. One may question whether one is exactly the right number; tournament results showed that this was dependent upon the environment it was in. In environments designed to exploit a willingness to forgive, being more forgiving did not fare as well as being less forgiving, while in an environment of cooperation, being more forgiving avoided mutual recrimination and did a little better.

Do Not Be Too Clever

Sophisticated programs did not do better than simple ones. Once again there is an important contrast between a zero-sum game like chess and a non-zero-sum game. In chess, it is useful to keep the other person guessing about your intentions. The more the other player is in doubt, the less efficient will be his or her strategy. This is useful where any inefficiency in the other player's behavior will be to your benefit. But in a non-zero-sum game, it does not always pay to be so clever. In the Quality Triad, you benefit from the other player's cooperation. Winning within the Quality Triad means making the rules clear and establishing the expectations of all concerned.

SUMMARY

Reacting to differences in expectations is necessary for a good working relationship. One strategy that has been successful in encouraging partnerships is a tit-for-tat strategy. This strategy (1) expects the best—it will not be the first to defect; it (2) is provocable—it retaliates for defection in the partnership, but after an immediate measured response it meets further expectations; it (3) is forgiving—it forgives differences when the partner meets expectations again.

Worksheet 17.1 gives a summary of the I.C.A.R.E. process through reciprocating.

Worksheet 17.1 A summary of the I.C.A.R.E. process through reciprocating.

Reciprocating Cooperation

Product/ Service Provided:_______

I. dentify Your Quality Triad

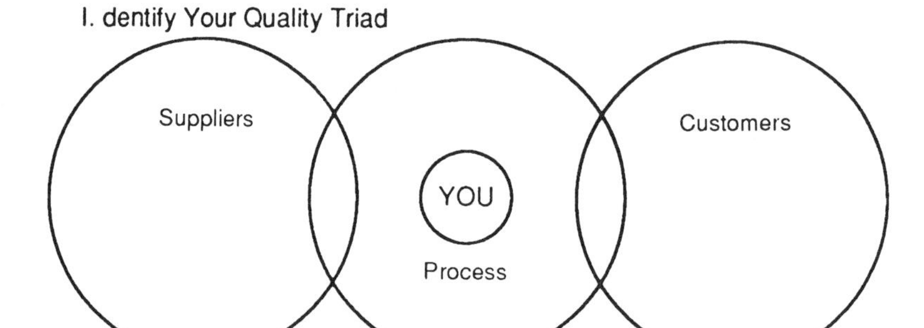

C. reate the Expectations

Assurance it is right	Self-control	Defined requirements
No supplier induced waste	No process induced waste	Feedback on product
Continual process improvement	No enduring problems	Suggestions for improvement

A. ccept the Responsibilities

Define wants	Work toward self-control	Assure it is right
Feedback on product	Reduce waste	Induce no waste
Suggest improvements	Continual process improvement	Continual process improvement

R. eciprocate Cooperation

Expect the best
Think win-win

Be provocable
Acknowledge cooperation/ noncooperation

Be forgiving
Let bygones be bygones

18

Enjoying the Benefits

A DUAL INTERPRETATION OF THE LINKAGES BETWEEN THE QUALITY TRIAD AND THE QUALITY SYSTEM

The linkage between the financial measures of quality and the Quality Triad started with the financial measures and flowed to the Quality Triad. The linkages connecting these aspects are shown in Figure 18.1.

Figure 18.1 reflects the history of quality systems. The first quality systems were born of conflict. They were developed to meet the customers' needs and minimize internal failure costs. Problems and headaches that threatened the smooth and unruffled course of production tended to be dominant concerns. Much of the growth of quality assurance has been the result of this conflict. For some companies, fire fighting has been replaced by fire prevention; yet trouble need not be the focus.

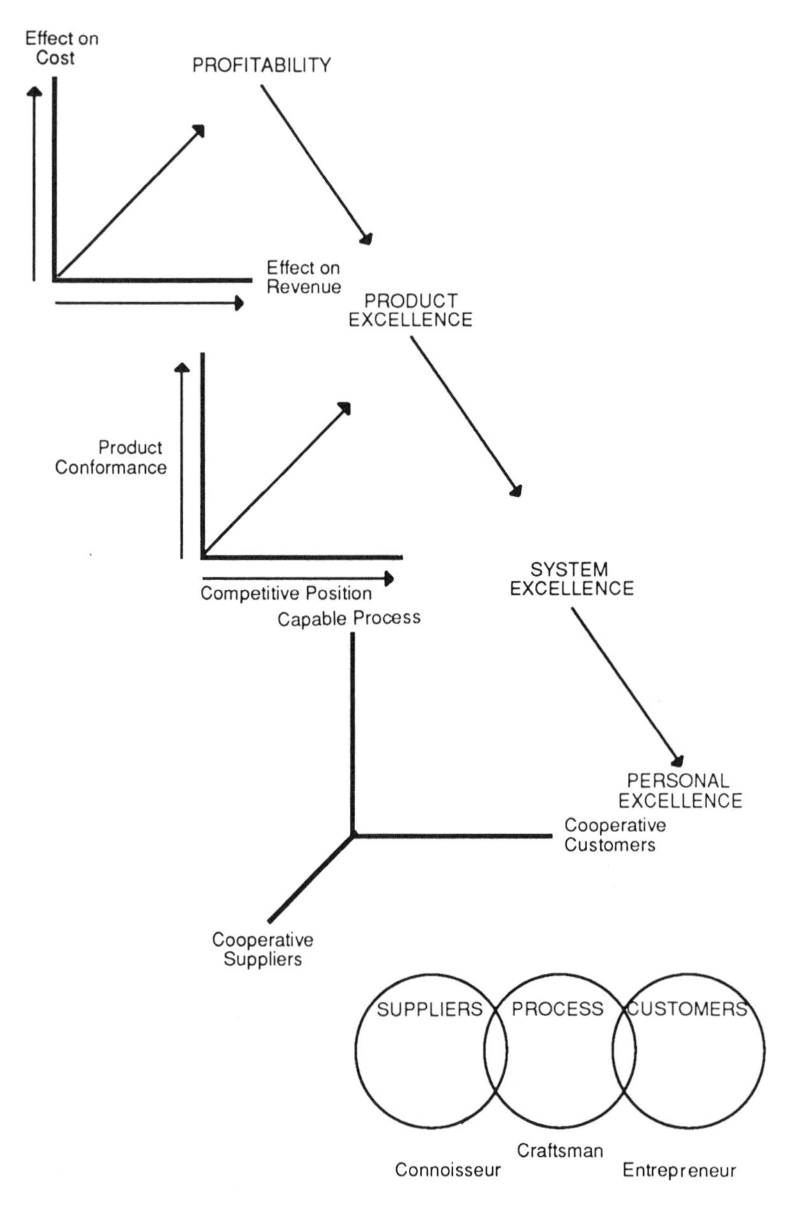

Figure 18.1 Quality linkages.

A complementary viewpoint can be seen in the Quality Triad. For the craftsman, the focus was seeing a beautiful way of doing the job; for the connoisseur, it was the appreciation of the subtleties in the product or service; for the entrepreneur, it was the building of a dream (building a cathedral, not just breaking rocks).

There is also a peace of mind associated with each role. The craftsman has mastery over a capable, stable process. He or she is confident that what is produced is right and right the first time.

In the relationship between the connoisseur and the entrepreneur, peace of mind is having constructive, mutually beneficial, harmonious partnerships. For the connoisseur, it is the assurance that there will be no problems, and for the entrepreneur, it is the assurance that problems will be quickly identified and discussed.

A dual interpretation of the linkages between the financial requirements and the Quality Triad also exists. It is also possible to start with the Quality Triad and flow backward to the financial requirements. In this interpretation, quality starts with the beliefs of each individual as shown in Figure 18.2. It starts with the beliefs that there is an ugly and a beautiful way of doing everything and that the working relationship between the buyer and the seller should be mutually beneficial, harmonious relationship.

The craftsman expects self-control and a capable, stable process. The entrepreneur and the connoisseur expect that their trading partners will give them whatever they need to do their jobs and will not knowingly introduce waste, complexity, ot instability into their process. These expectations act as control limits. They trigger action and carry along with them matching responsibilities. These responsibilities and expectations provide activities for each role that result in quality products that can be measured financially.

This flow as reflected in Figure 18.2 also reflects the reality that all meaningful and lasting change starts from the inside and works its way out. It starts with the mind's perception of the world, with beliefs, and interprets what is happening in terms of these beliefs. Change imposed from the outside can be met with reluctance and resistance.

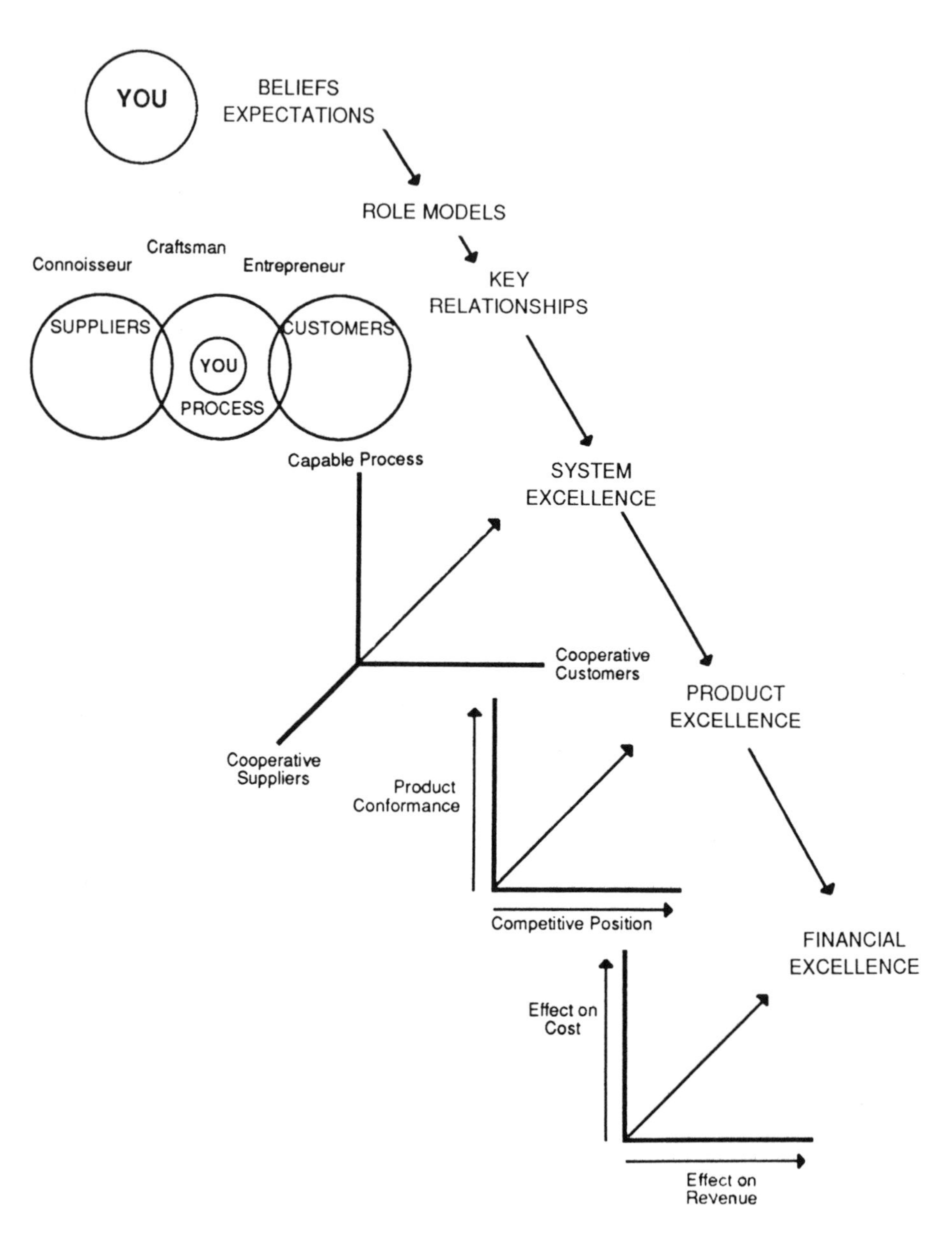

Figure 18.2 A complementary view of the quality linkages.

A REAL UNDERSTANDING OF QUALITY

Most quality systems are set up with the purpose of either reducing scrap and rework or increasing customer satisfaction. Because of the many linkages between the system requirement for financial performance and personal excellence, however, and because of the win/win nature of these linkages, there are several interpretations of the purposes of quality systems.

To illustrate this more fully, consider a typical element in a quality system: certification. Certification is the process of "licensing" operators prior to allowing them to work. Examples of jobs on which operators could be certified are soldering, welding, testing, and so on. The process normally consists of a training period during which the operator is instructed in the operation and then tested on the material covered. If the test score is acceptable, the operator is certified. If the operator fails the test, the operator will not be allowed to work, and further training will be required.

Thus, the operator is prevented from producing products that require significant scrapping or reworking, and nonconforming products are not sent on to the customer. So the financial system is served.

The quality master has a slightly different understanding of this. He or she realizes that quality does not just serve the system but can also be put to his or her personal use. With this understanding, the quality master realizes that quality is a win/win situation among many linked outcomes. The quality master starts with the belief that there is a beautiful way of doing everything. This belief generates the expectation of self-control.

To meet this expectation, training and feedback on training must occur. The quality master realizes that violating this expectation not only detracts from personal excellence and craftsmanship but also can result in a loss as measured in the financial system, a decrease in product excellence, and a loss of credibility with a key customer. The quality master realizes that all these aspects are connected. In fact, the craftsman can use this as a selling point in the justification of the necessary training.

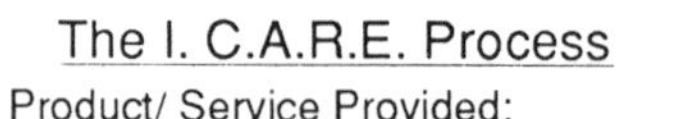

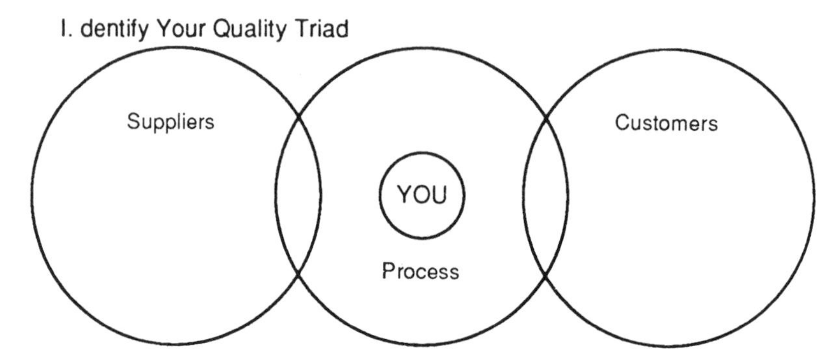

C. reate the Expectations

Assurance it is right	Self-control	Defined requirements
No supplier induced waste	No process induced waste	Feedback on product
Continual process improvement	No enduring problems	Suggestions for improvement

A. ccept the Responsibilities

Define wants	Work toward self-control	Assure it is right
Feedback on product	Reduce waste	Induce no waste
Suggest improvements	Continual process improvement	Continual process improvement

R. eciprocate Cooperation

Expect the best
Be provocable
Be forgiving

E. njoy the Benefits

"A real understanding of quality captures the system and puts it into personal use."

Figure 18.3 A complete summary of the I.C.A.R.E. process.

A similar situation occurs in almost all the elements of a quality system. Quality circles and participative team solving may be ways to decrease failure costs, but they can also be viewed as means of eliminating enduring problems. Incoming inspection or gate inspections are installed to identify problems prior to making further substantial investment in the product, but they can also be viewed as means of providing feedback for self-control or as means of assurance that the material will work correctly. Similarly, statistical process control can be viewed as a tool to make uniform, consistent products time after time, but the craftsman could envision this as a listening tool helping him or her stay in touch with the process and maintain a mastery of it. A complete summary of the I.C.A.R.E. process is shown in Figure 18.3.

Conclusion: A Starting Point for Quality

I have emphasized how the Quality Triad complements a top-driven quality program. The best of all would be where both exist, where active top management participates in a company-wide quality program supported by individual craftsmen, connoisseurs, and entrepreneurs.

But what if the company you work for has not yet been enlightened about quality? Some may say that all is lost. There has been much written about company-wide quality programs that have failed because of a lack of management support.

I will be the first to say that this situation is not easy; but people may perceive false limitations on what can be done. Improvements can be made within the Quality Triad, though, this will not change a whole company. It will, however, let you operate with excellence and affect your process, your suppliers, and your customers. You may not be able to change the whole world, but you can change your own working world.

Most of the aspects can even be "secretly" pursued, because the roles, responsibilities, and expectations are determined almost on an individual level. Although they can be secretly pursued, they do not have to be. Each activity supports the win/win relationship between the individual and the company.

The key in this situation, however, is not what management is going to do or not do. The key is deciding how you will interact with your process, your customers, and your suppliers. The choice is yours.

Index